200 - 2D 3D CAD EXERCISES

Kovalan.S

200 - 2D 3D CAD EXERCISES

Copyright © Kovalan Sandiyappan, 2020.

No part of this book may be reproduced in a retrieval system or transmitted in any form or by any means electronics, mechanical, photocopying, recording and or without the permission of the author.

    Author   : Kovalan Sandiyappan.

    Publisher : Kovalan Sandiyappan.

    Email   : kovalans@outlook.com

Table of Contents

1. Preface   4
2. Dedication   5
3. Disclaimer   6
4. Resources & link to bonus CAD book.   7
5. 100 no's - 2D CAD Exercises   8
6. 100 no's – 3D CAD Exercises   60

Preface

This book was created to feed budding CAD designers with a variety of 2D and 3D CAD Exercises while they are learning and mastering the art of CAD.

This Work Book is a Collection of the Best drawings from the original 2D 3D CAD EXERCISES Volume 1, 2 & 3.

This book is a CAD neutral Exercise book and can be used along to, practice and learn leading 2D and 3D CAD modelling software out there in the market.

The exercises have been arranged in order of easiness to difficulty letting you practise from basic commands to advanced commands. This book is not a tutorial to teach CAD, but is only to practise with the contained exercise drawings. All the drawings have been dimensioned to facilitate the reader to recreate them.

Dedication

I dedicate this book to

My Family

&

Friends.

A special thanks to
Nirmal.J & Sasidharan.M
for their valuable support.

Disclaimer

In this book you will find 100 no's - 2D CAD Exercises and 100 no's - 3D CAD Exercises to practice for students and others learning a 3D CAD Modelling software. The exercises have been arranged from easy to hard, and progressively arranged to include most of the commands you would find on a 2D / 3D CAD modelling software.

All the drawings have been dimensioned to facilitate easy replication of the drawings.

This book does not contain any step by step instructions to use any software or to draw any of the exercises. Readers will have to use their knowledge of the software to reconstruct the drawings with the dimensions provided.

Resources

Additional 2D & 3D CAD Exercise book can be downloaded from the following web site: www.2d3dcadex.com

Additional practice exercises may also be made available from time to time in the above web site.

Most leading CAD software makers offer trial / student versions. For trial version installation of 2D or 3D CAD Modelling software search in www.google.com with search string like:<CAD software name> trial

You could find links to download your desired software trial version with which you can practice the CAD Exercises. You may also purchase the Student version of the software from their official web site.

2D CAD EXERCISES

Page intentionally left blank

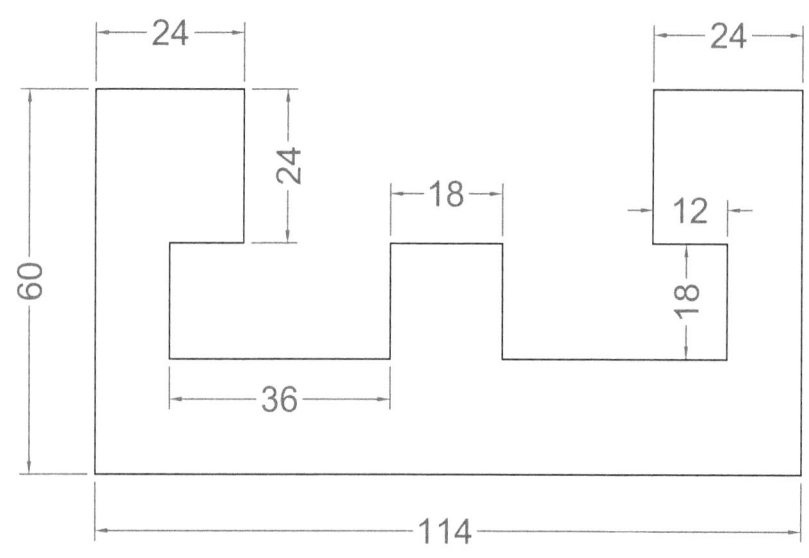
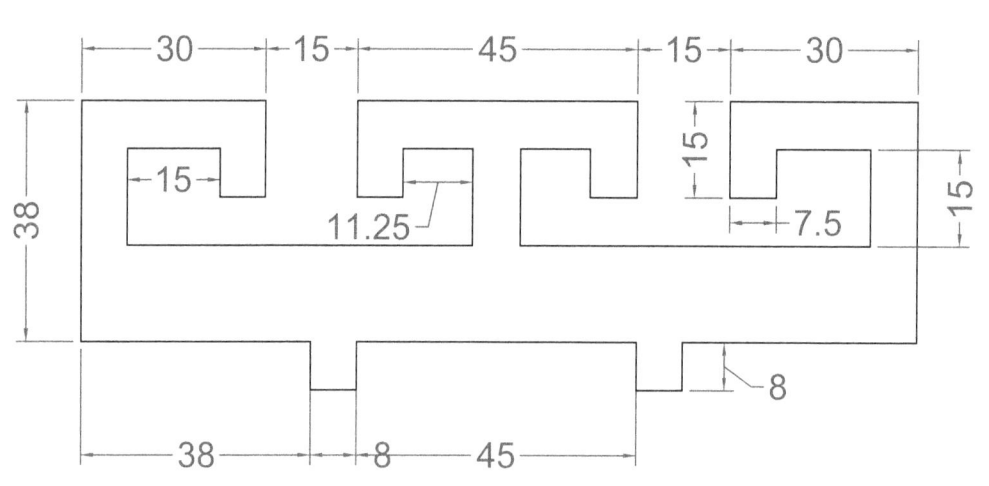

ALL DIMENSIONS ARE IN MM

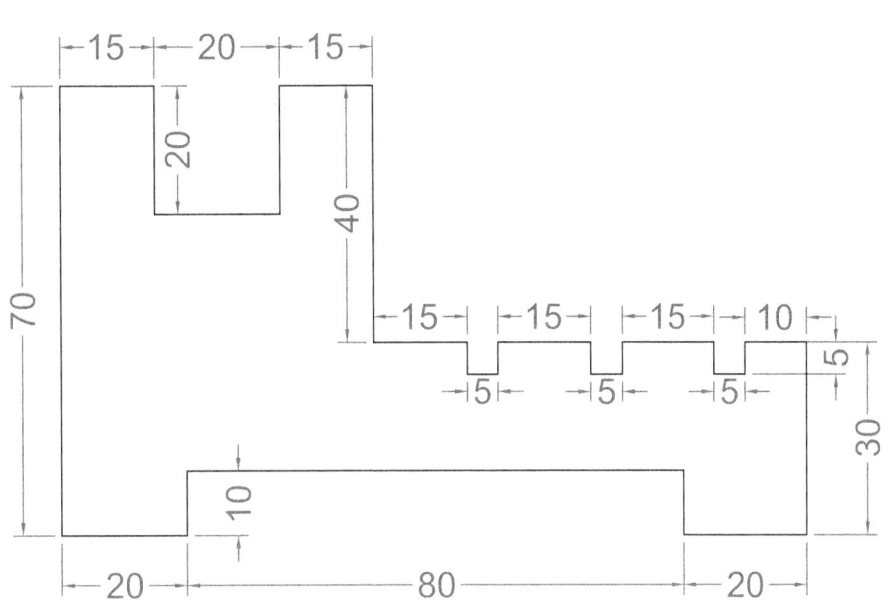

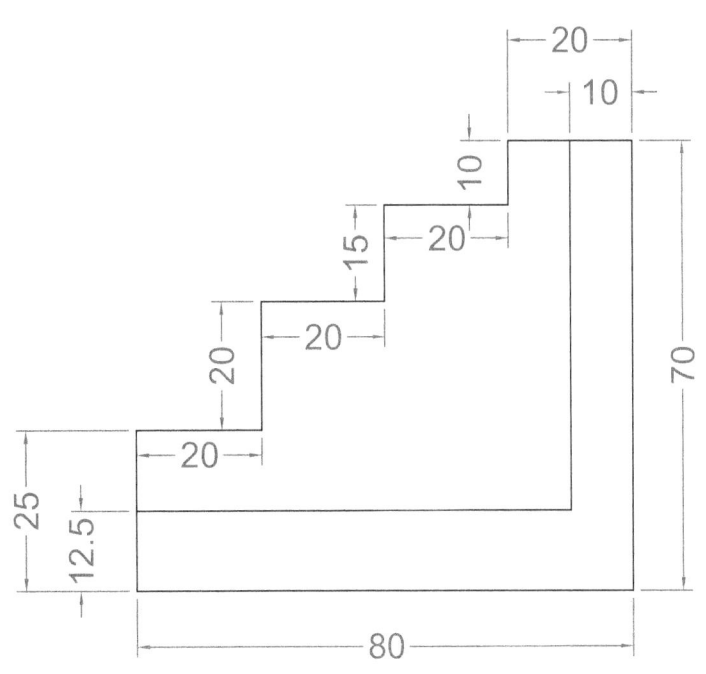

ALL DIMENSIONS ARE IN MM

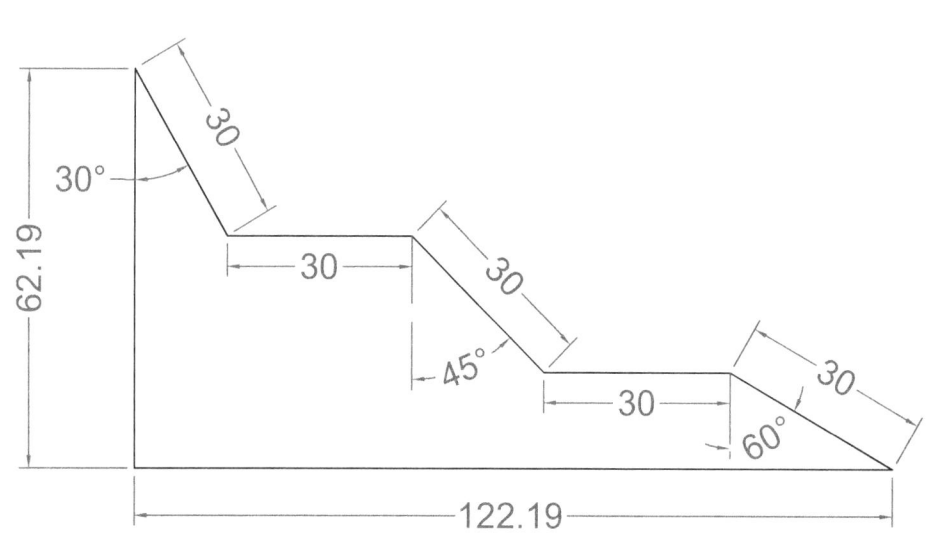
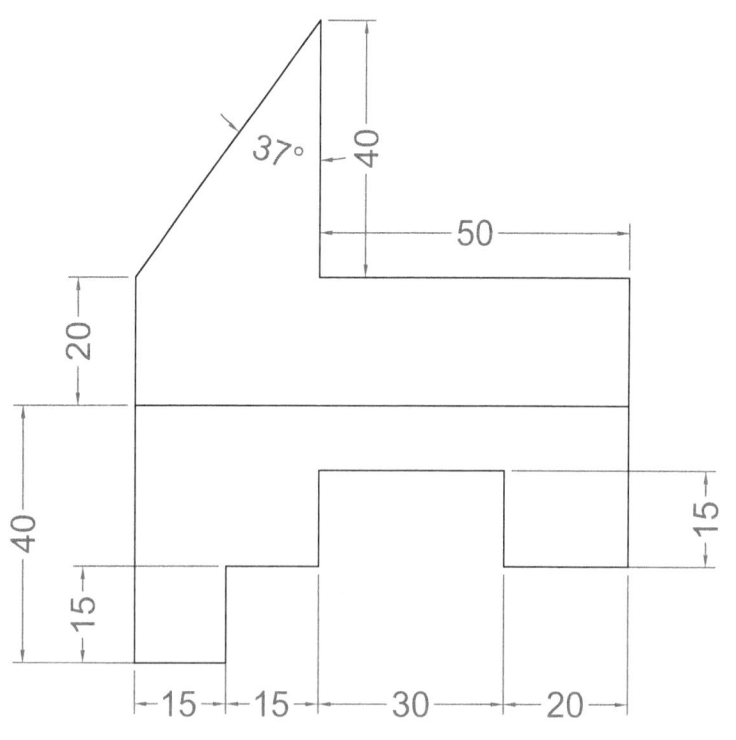

ALL DIMENSIONS ARE IN MM

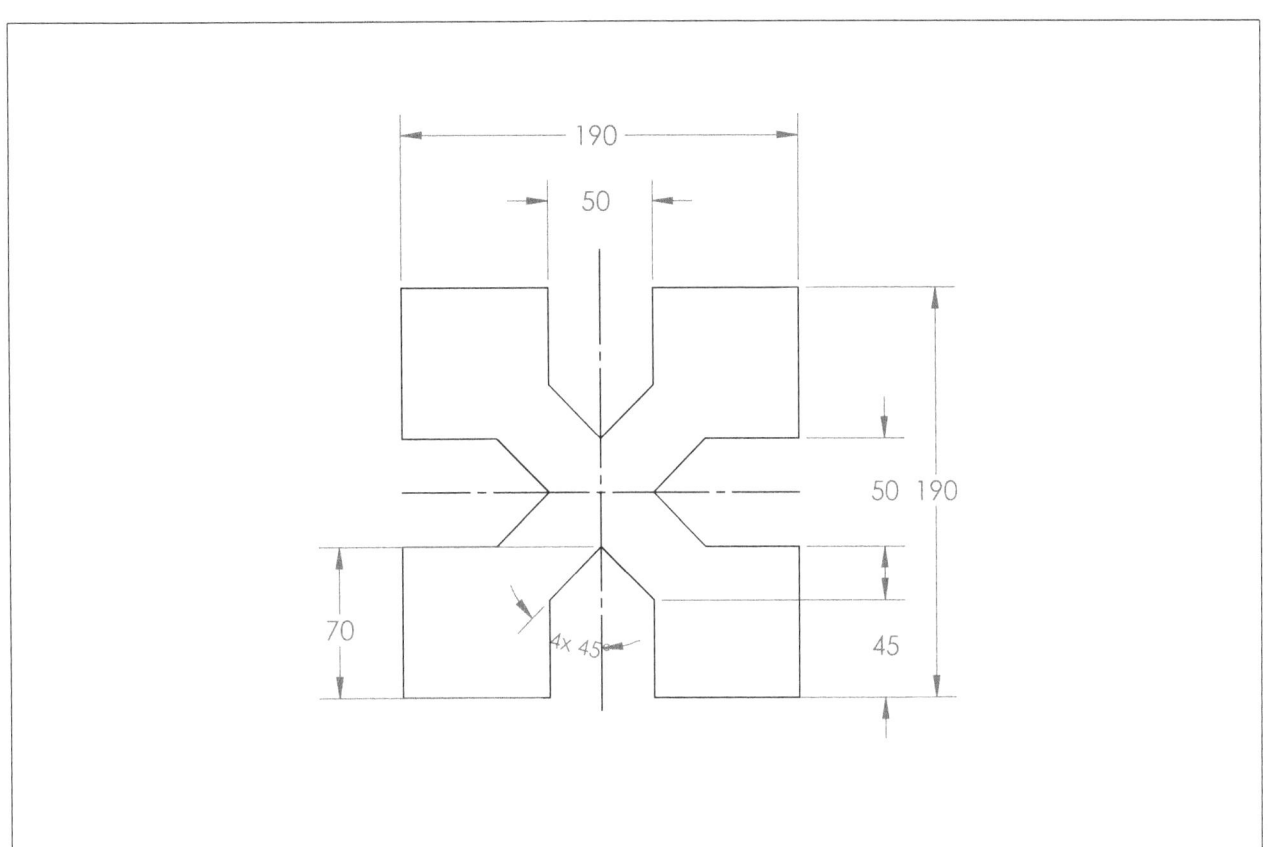
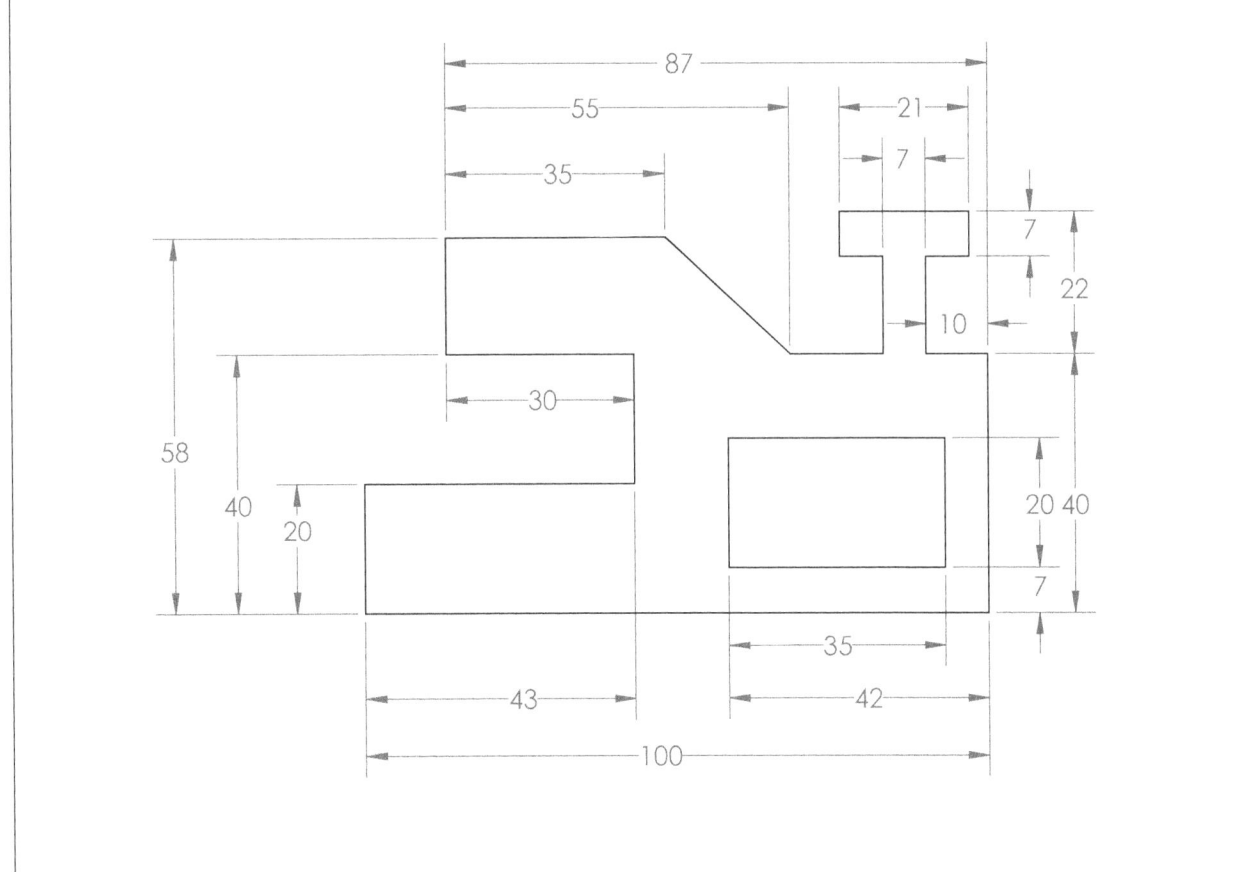

ALL DIMENSIONS ARE IN MM

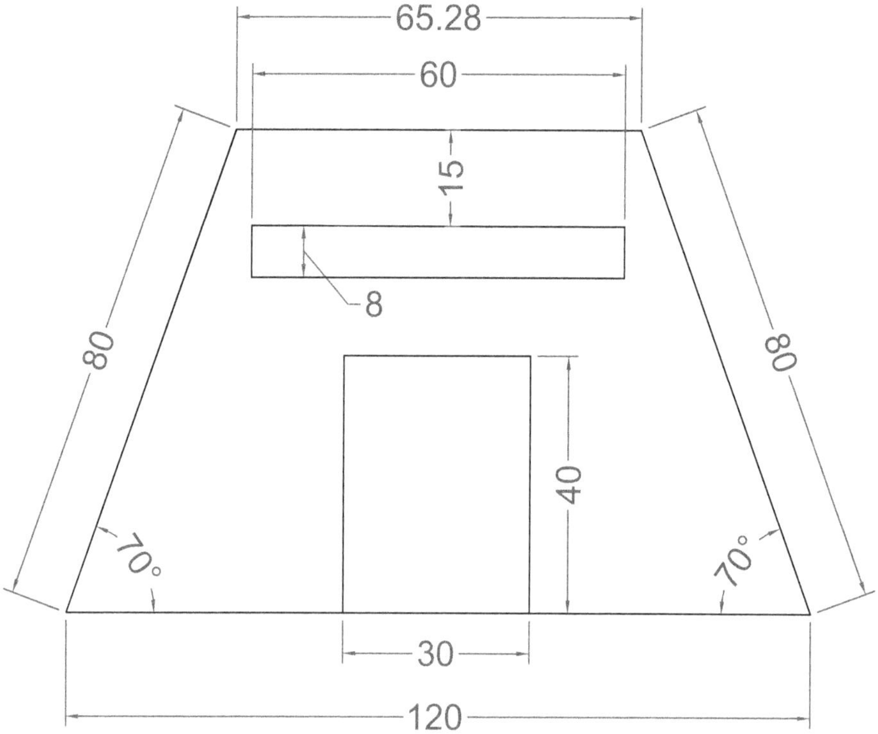

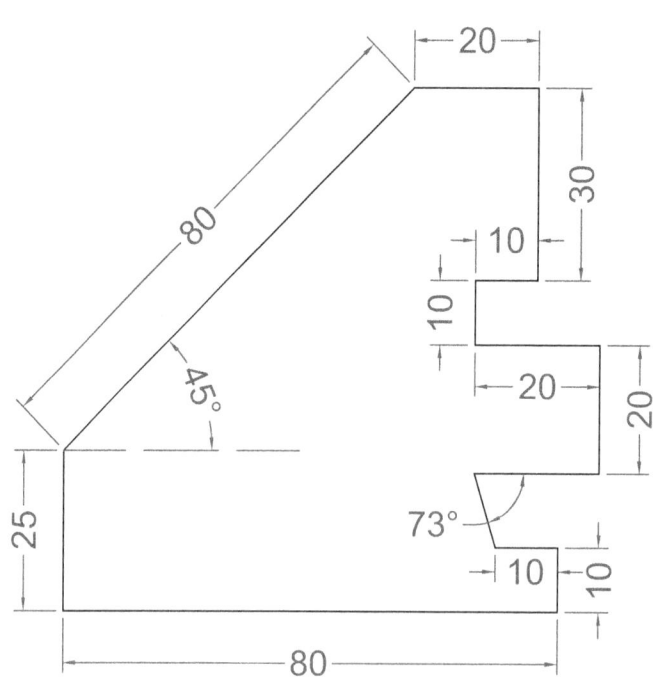

ALL DIMENSIONS ARE IN MM

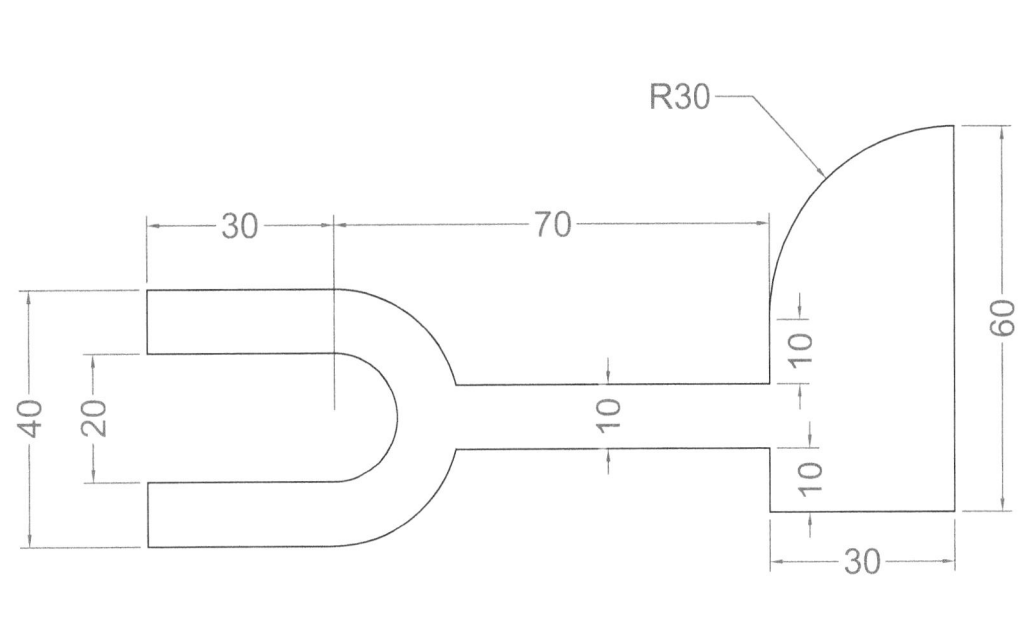
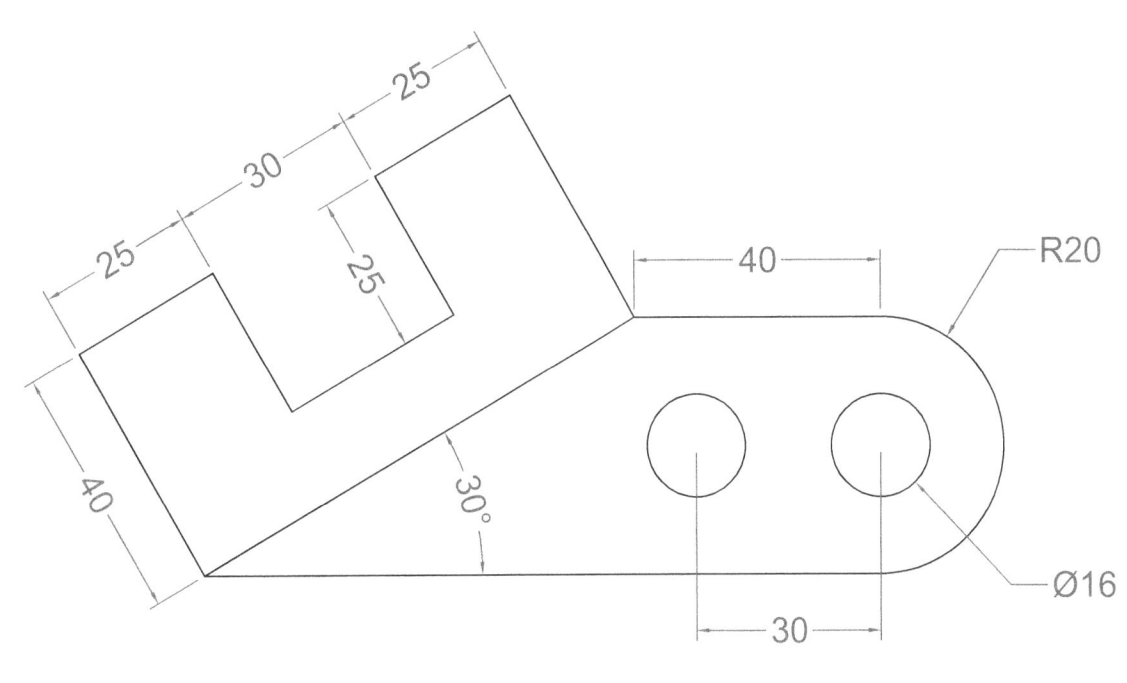

ALL DIMENSIONS ARE IN MM

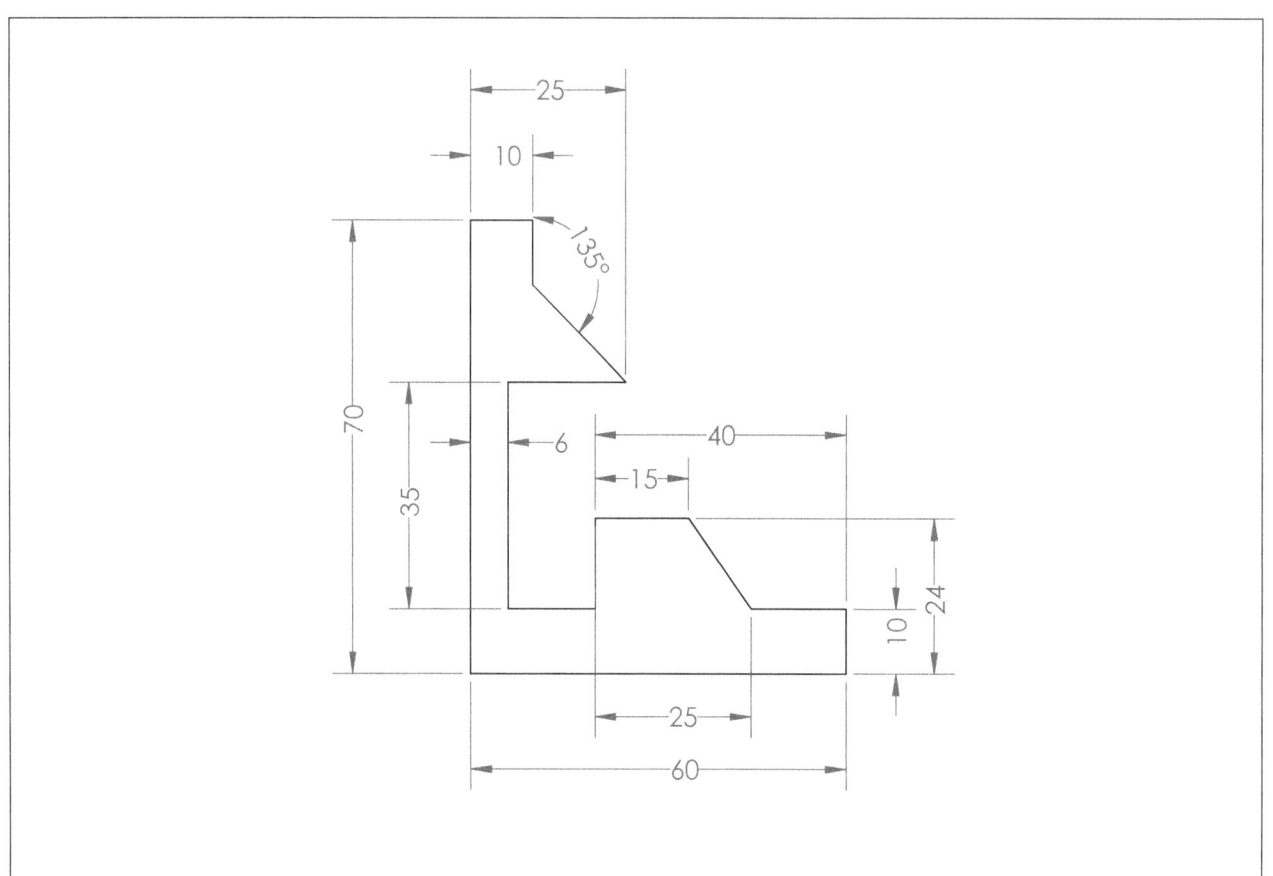
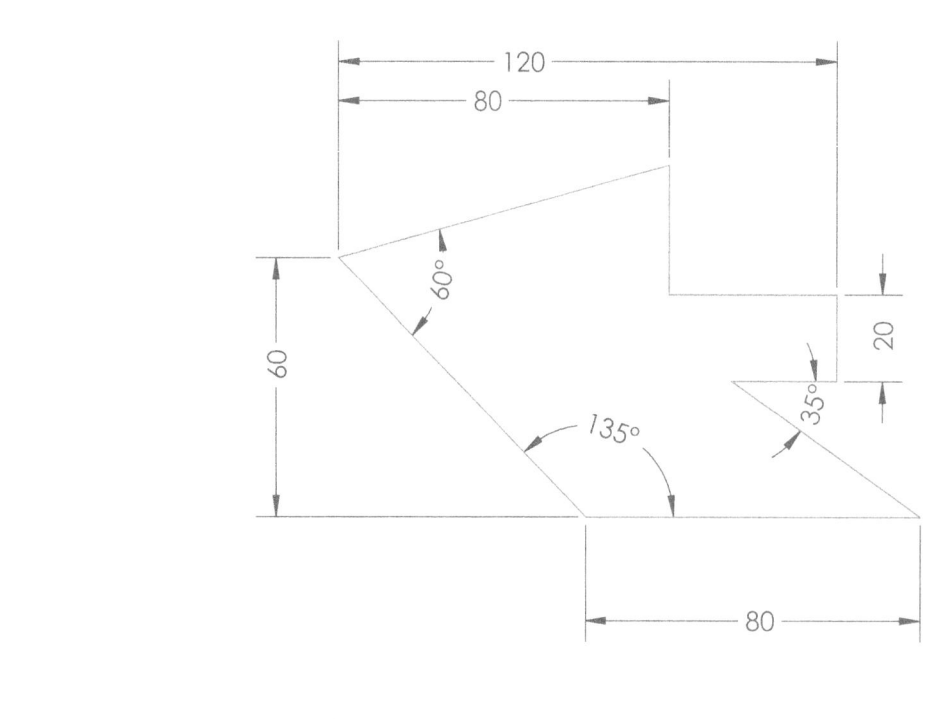

ALL DIMENSIONS ARE IN MM

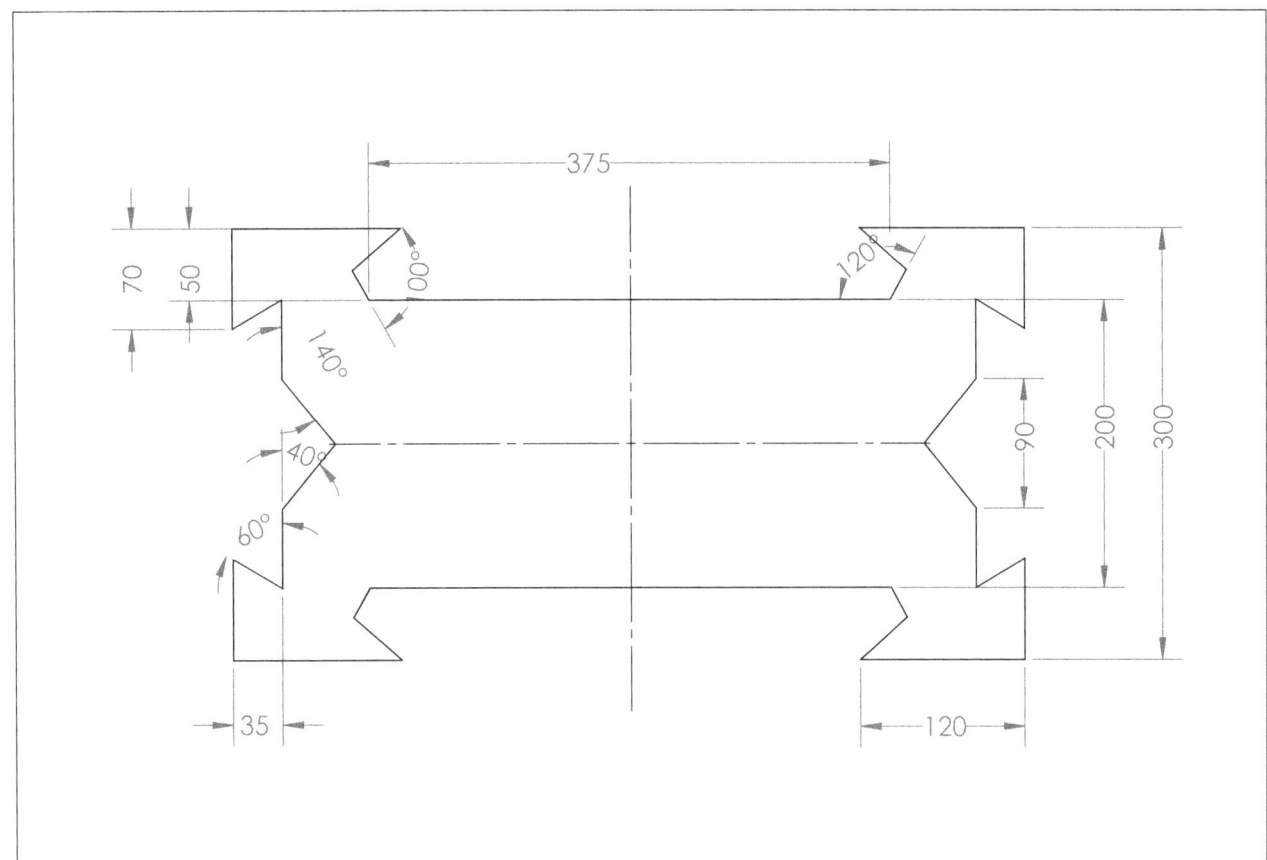
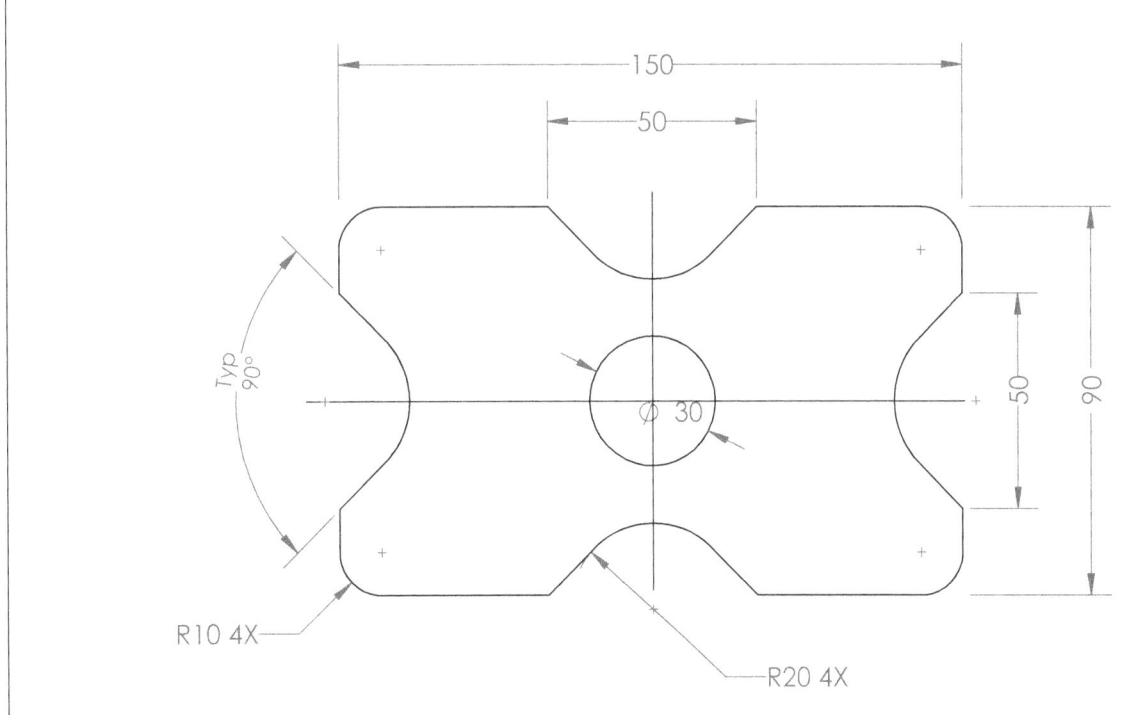

ALL DIMENSIONS ARE IN MM

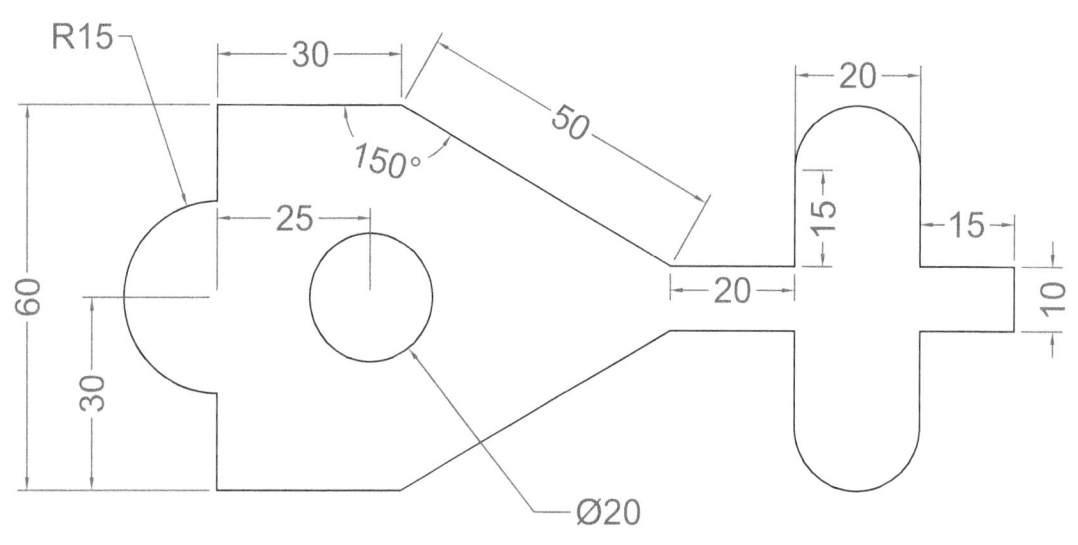

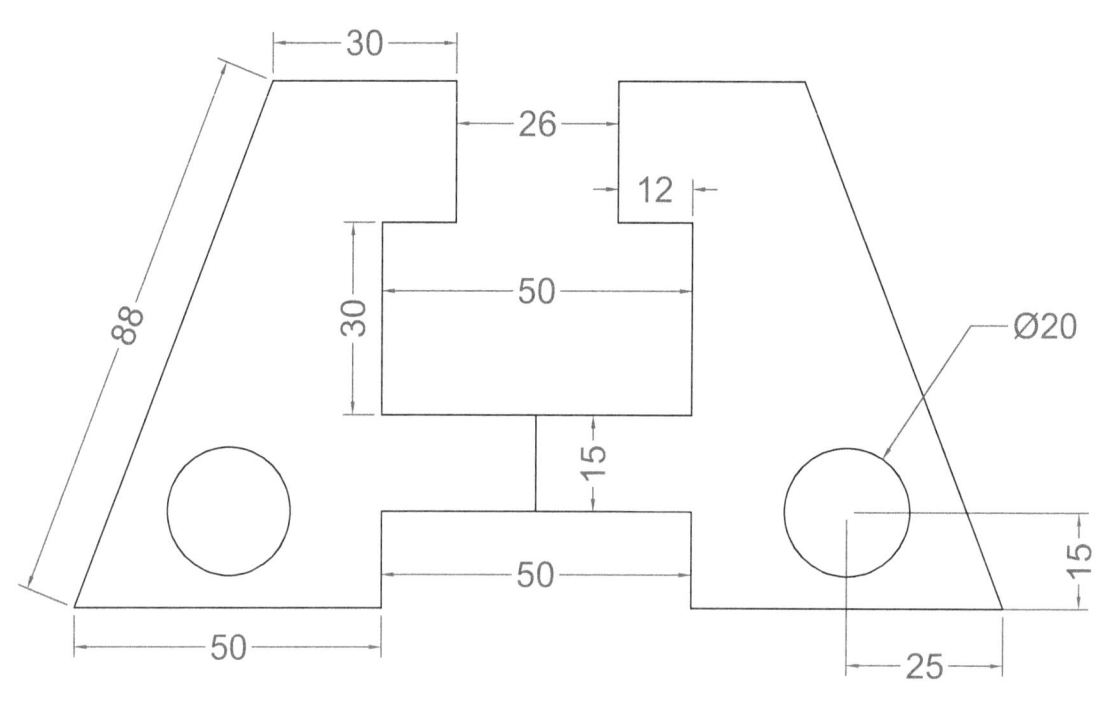

ALL DIMENSIONS ARE IN MM

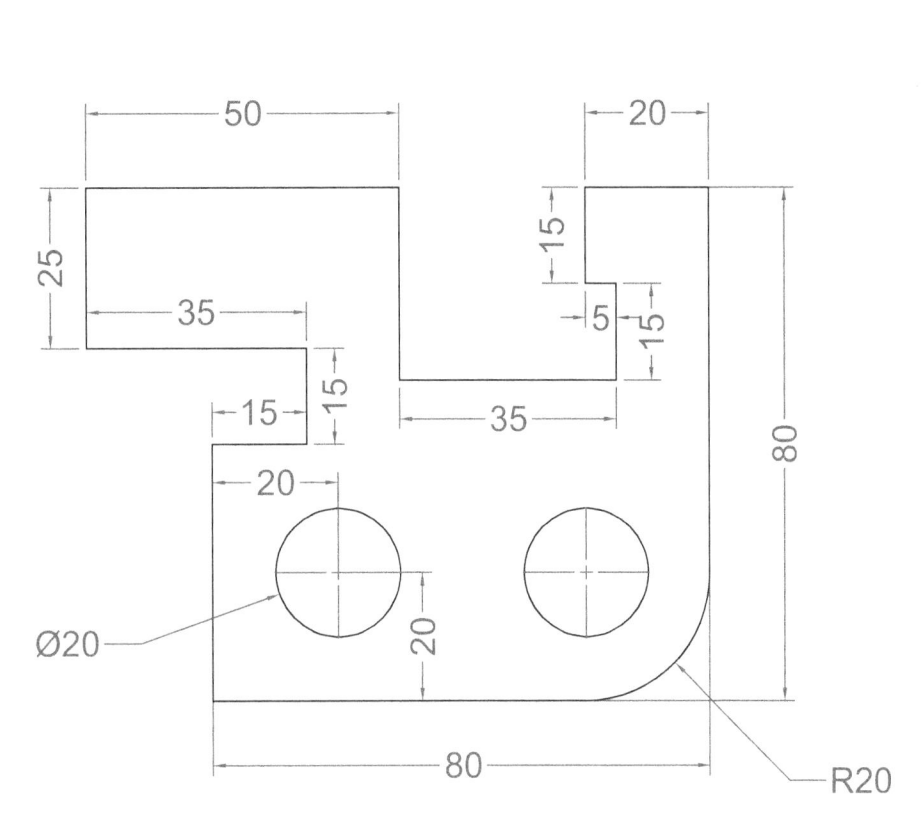
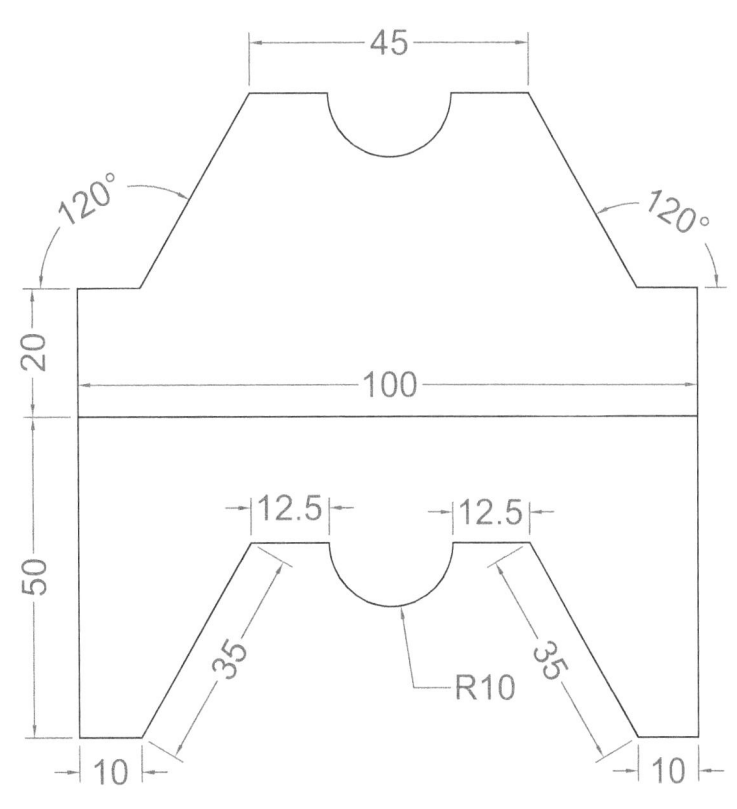

ALL DIMENSIONS ARE IN MM

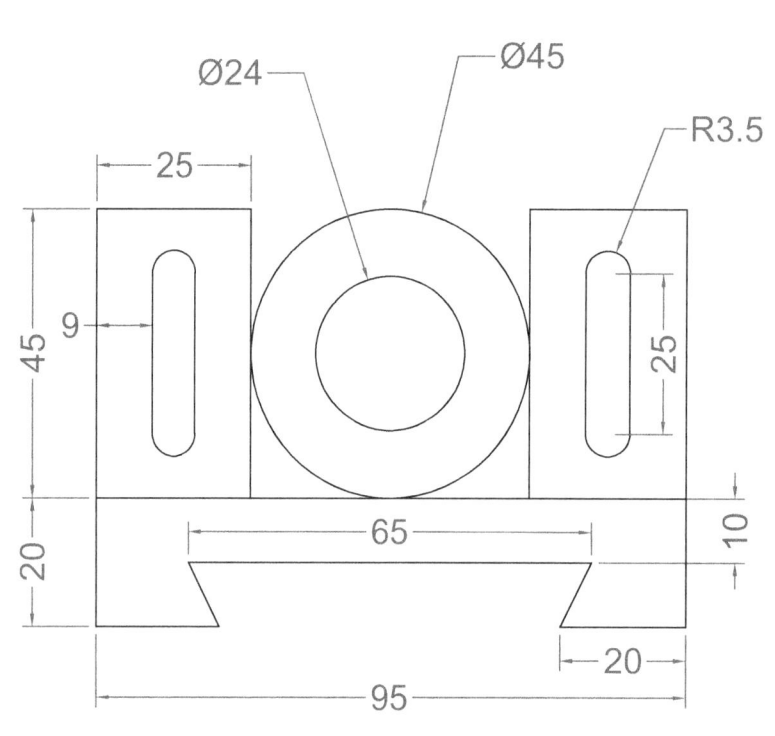
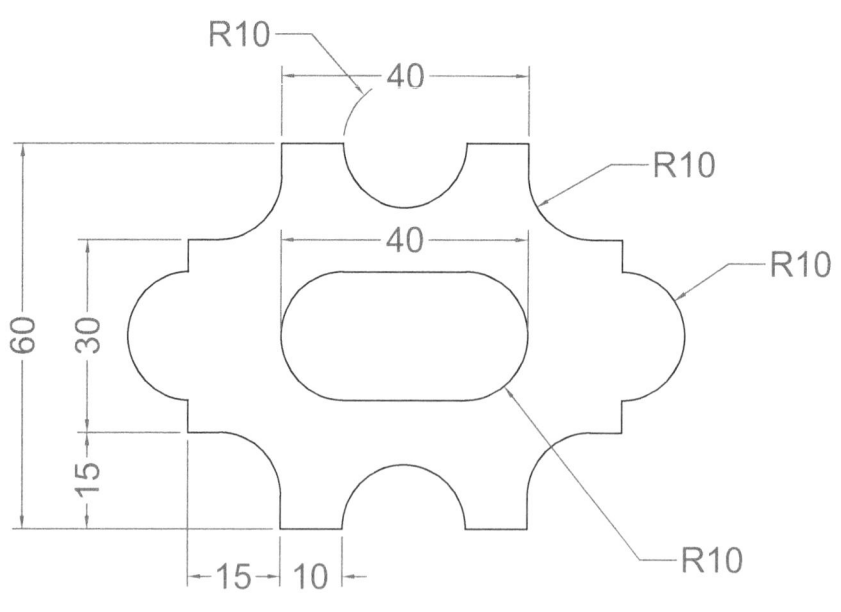

ALL DIMENSIONS ARE IN MM

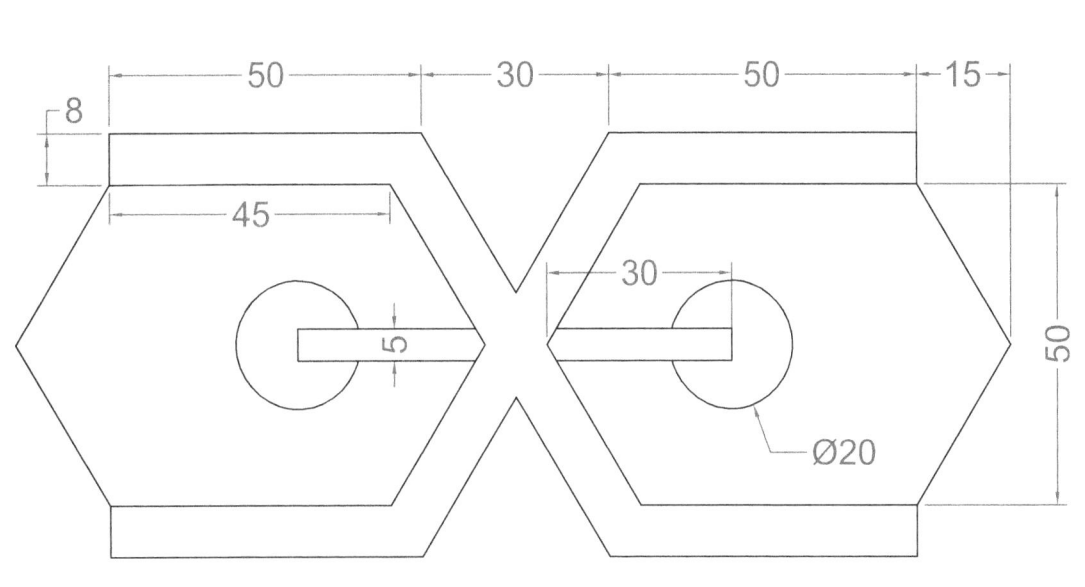

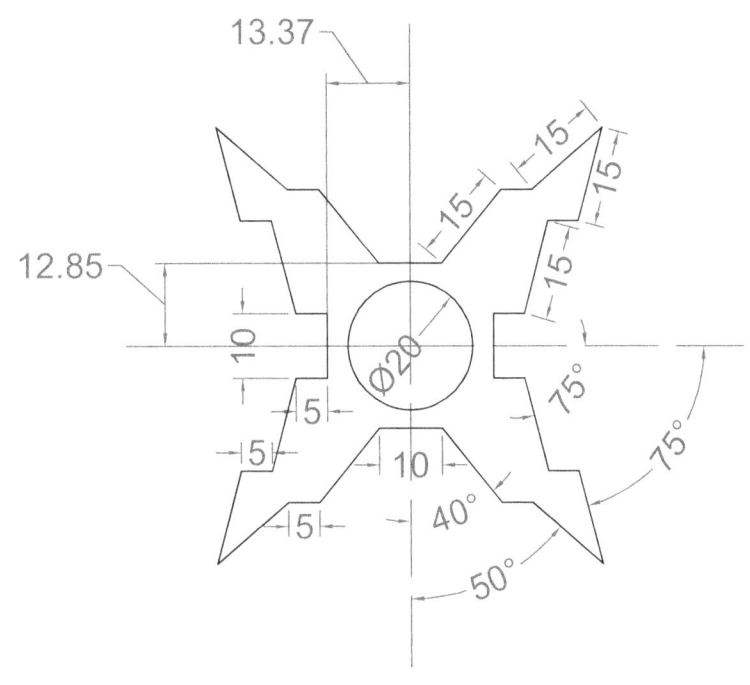

ALL DIMENSIONS ARE IN MM

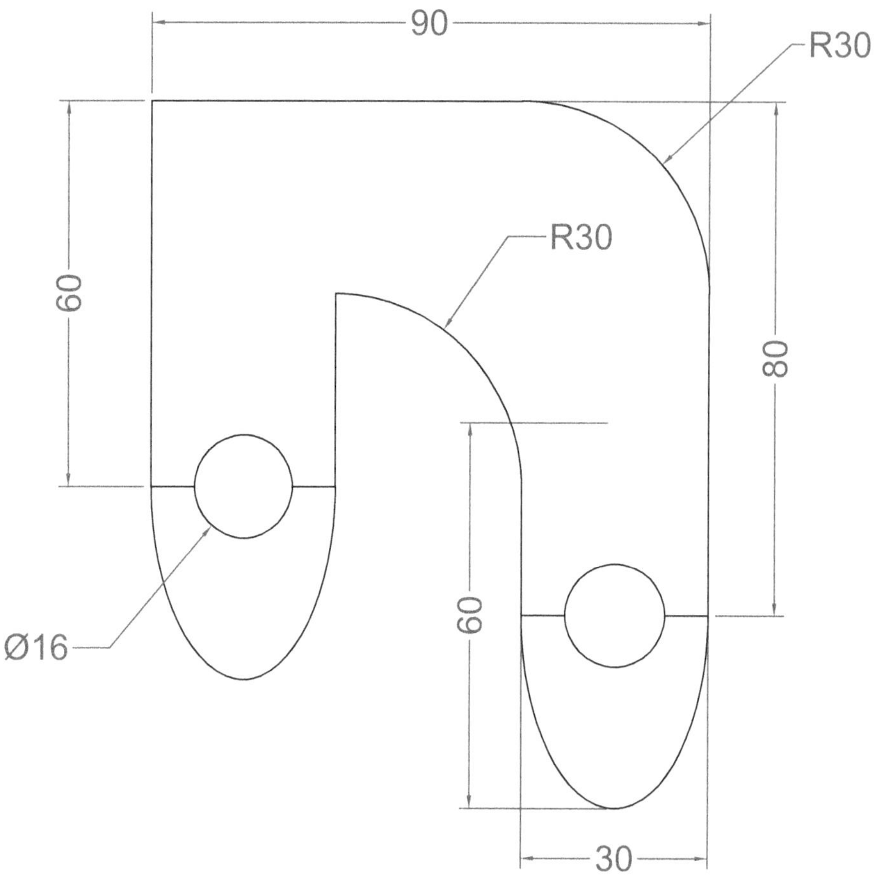

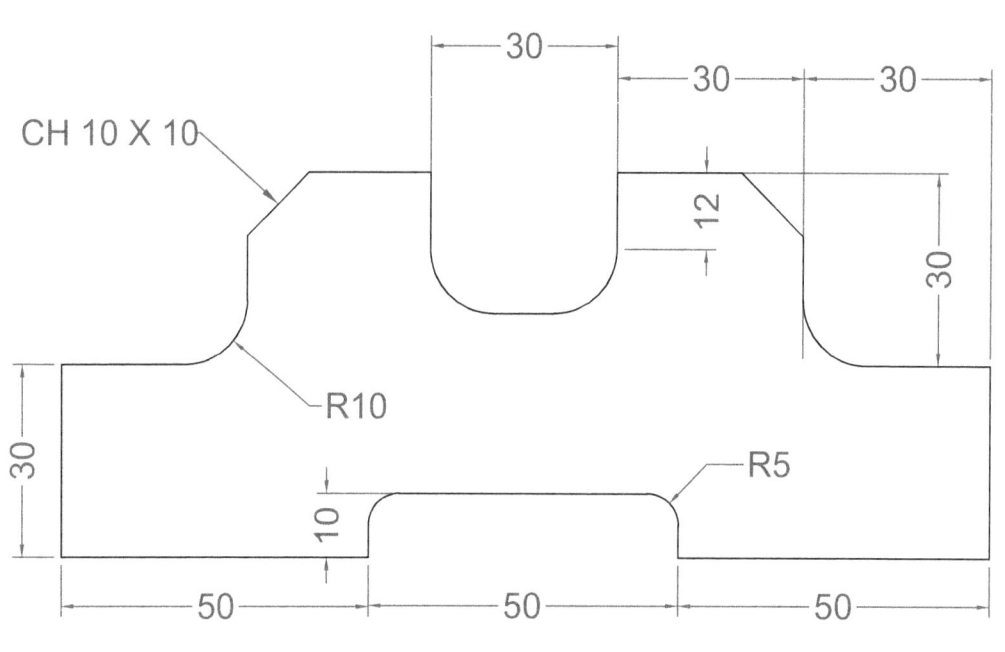

ALL DIMENSIONS ARE IN MM

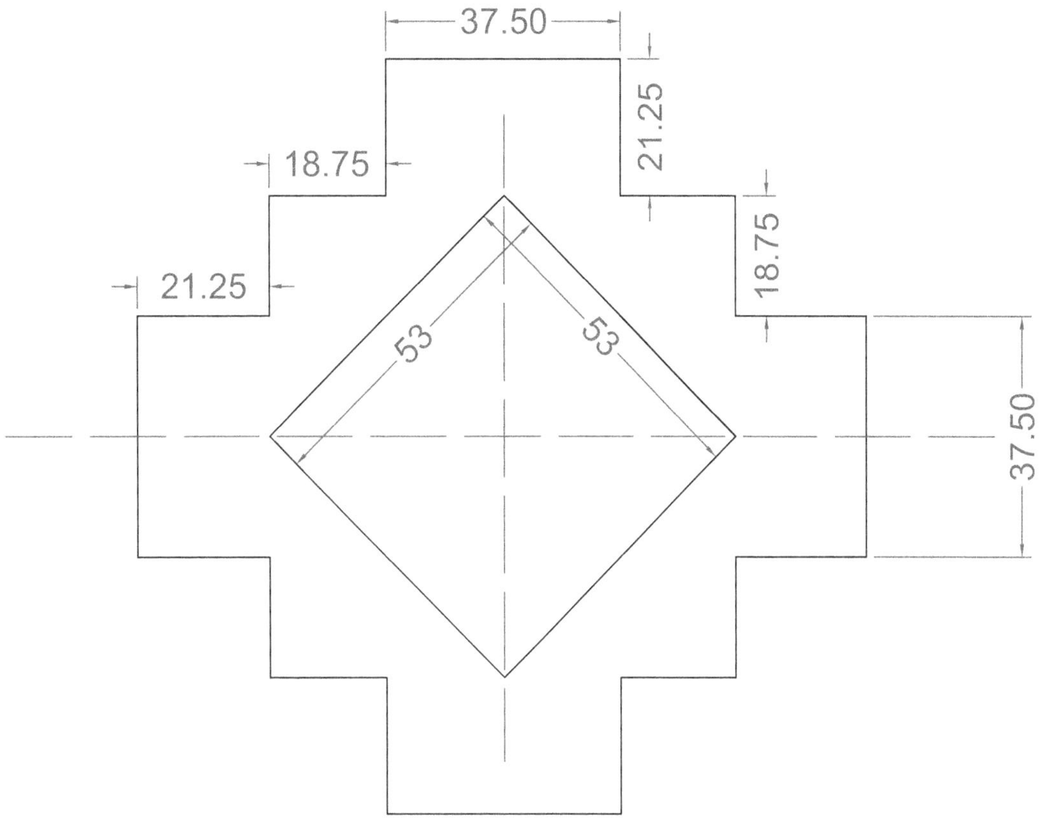

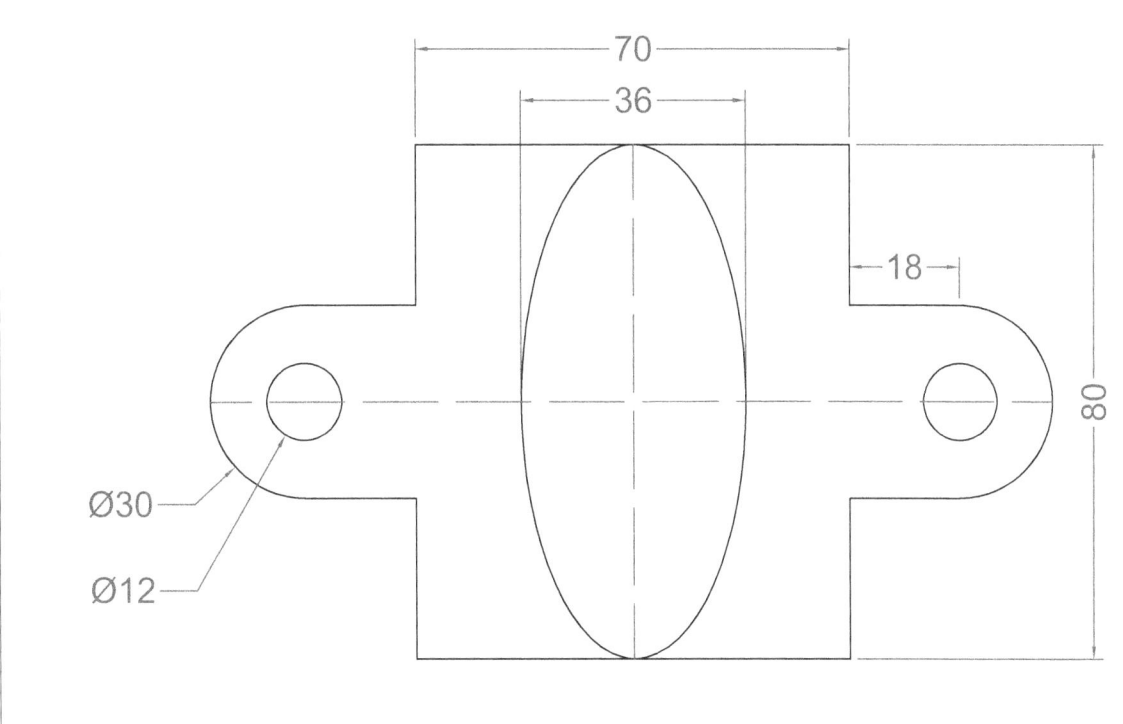

ALL DIMENSIONS ARE IN MM

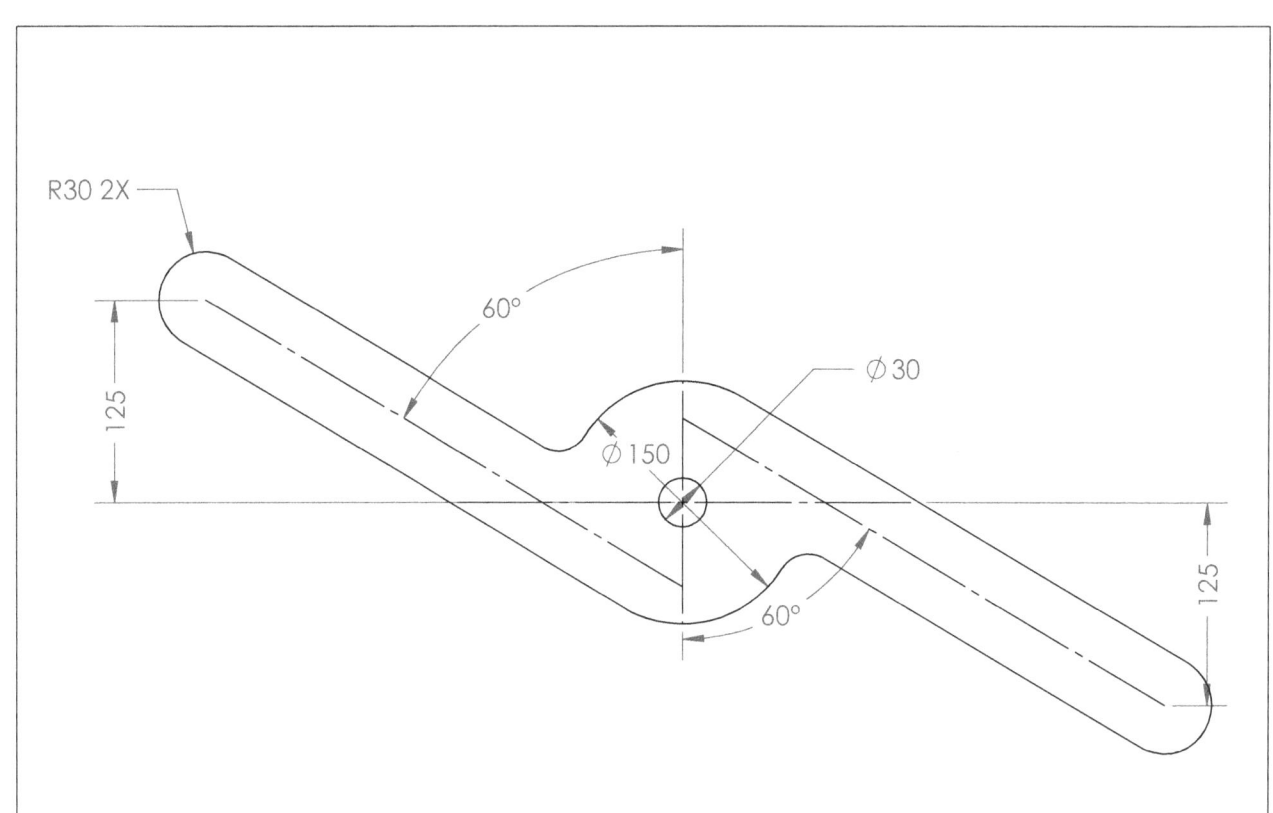
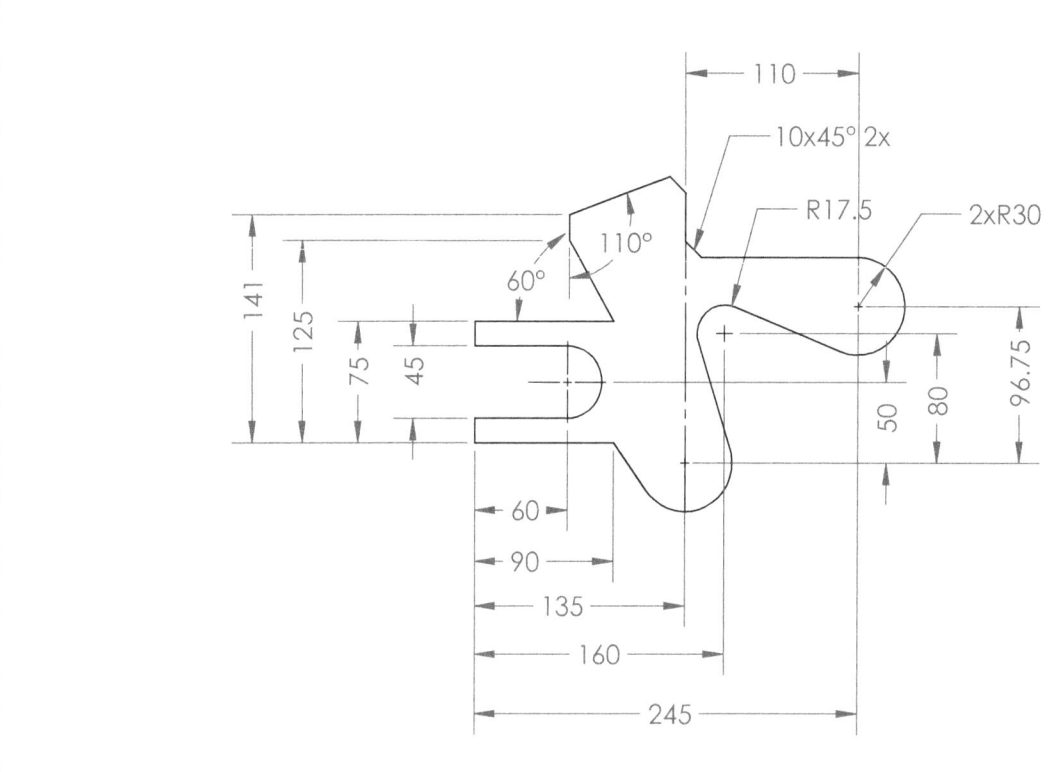

ALL DIMENSIONS ARE IN MM

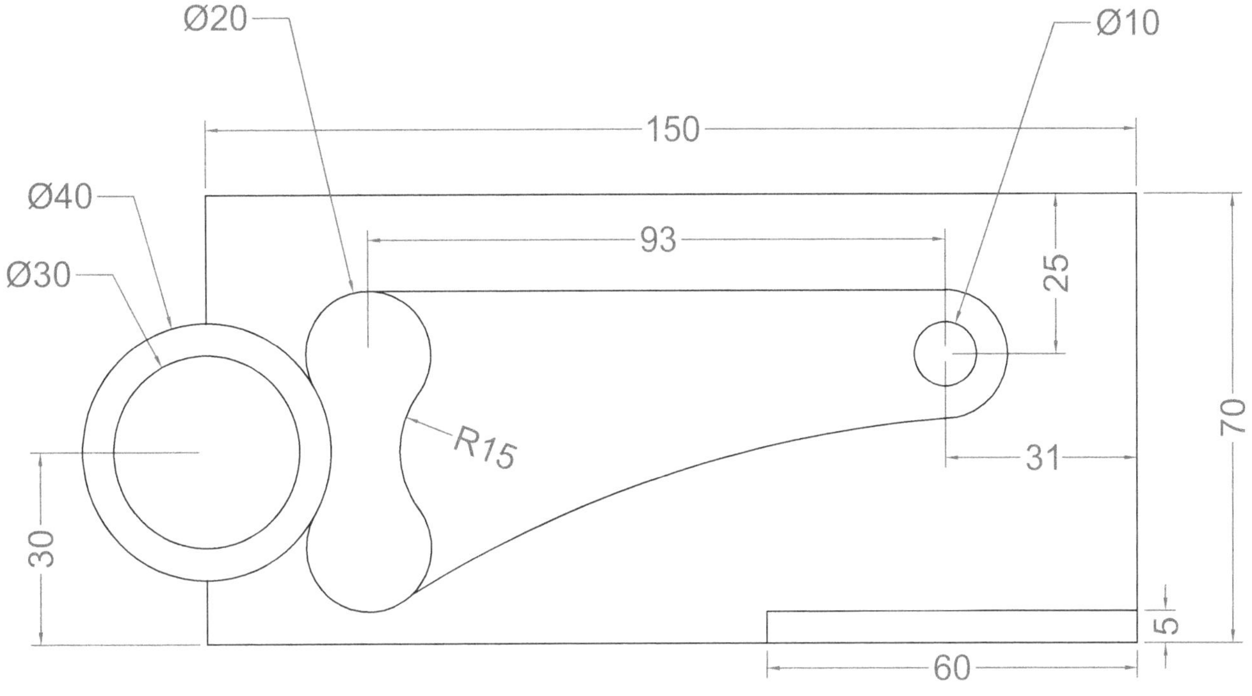

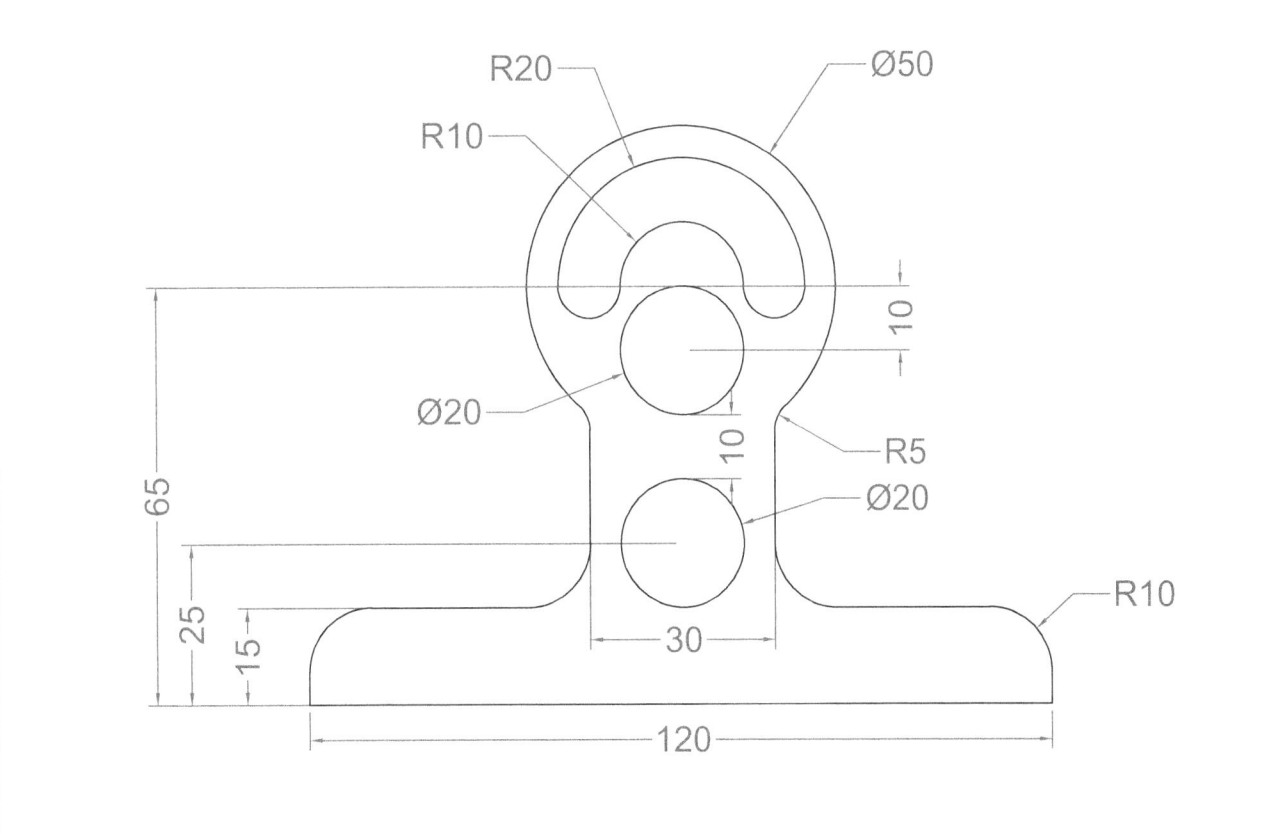

ALL DIMENSIONS ARE IN MM

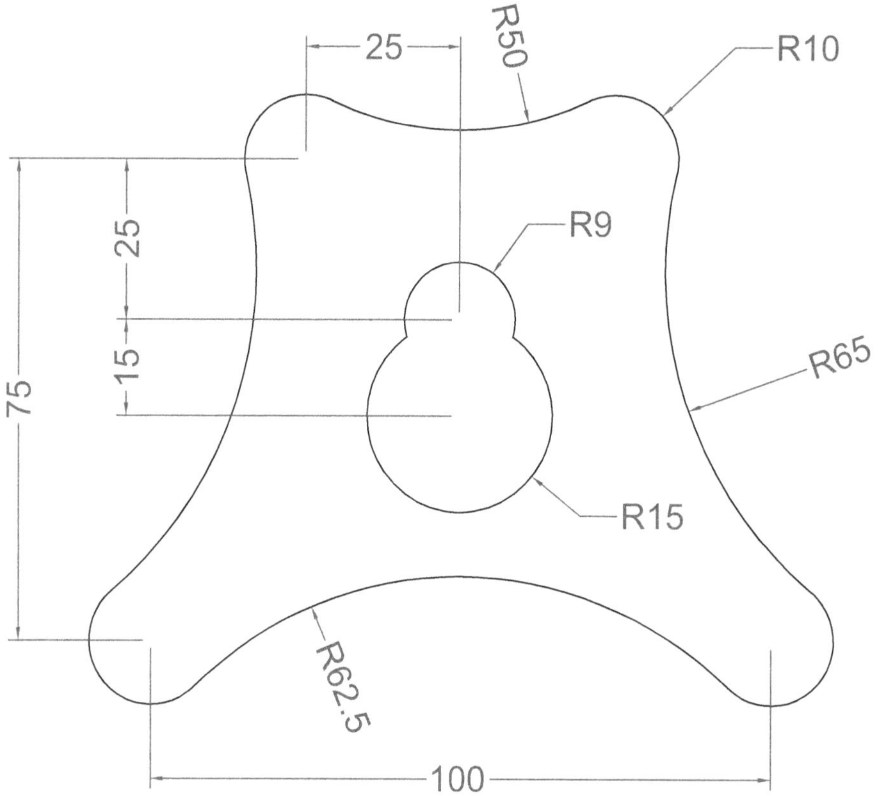

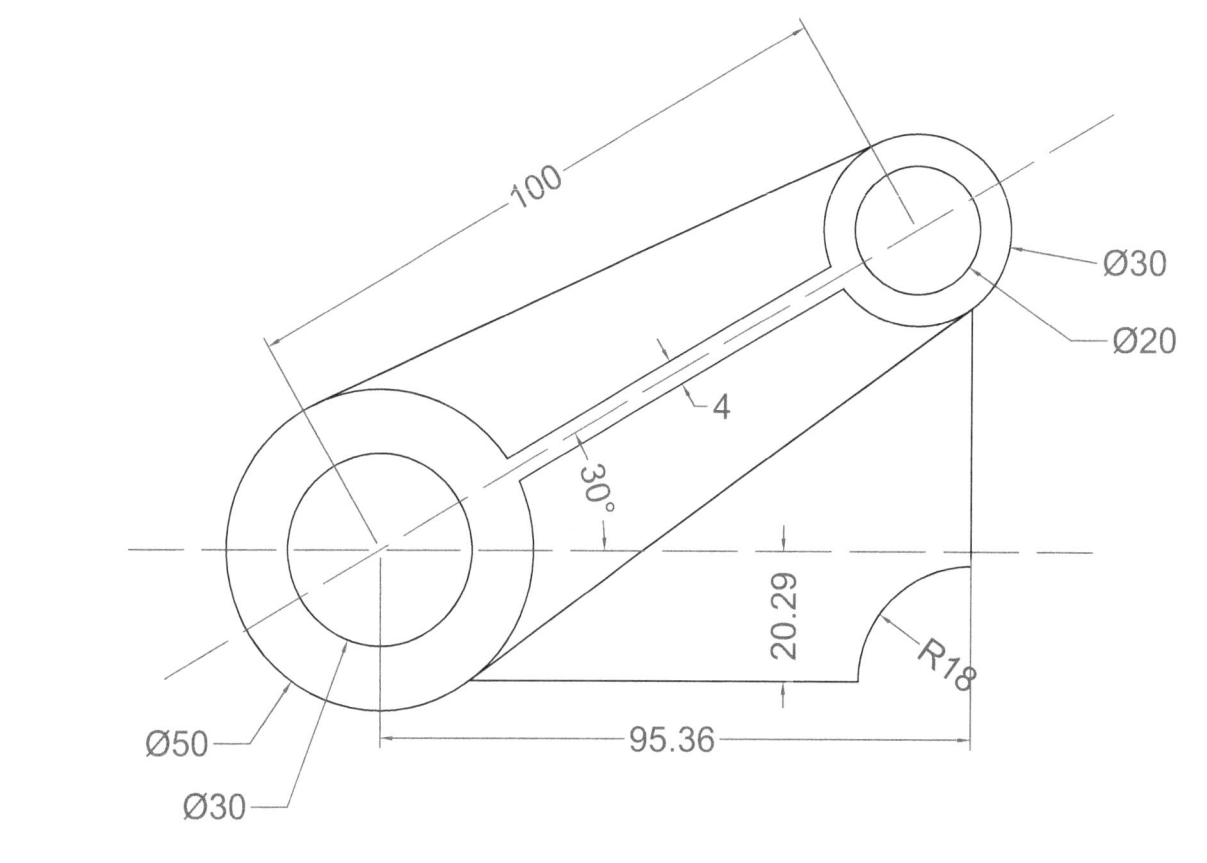

ALL DIMENSIONS ARE IN MM

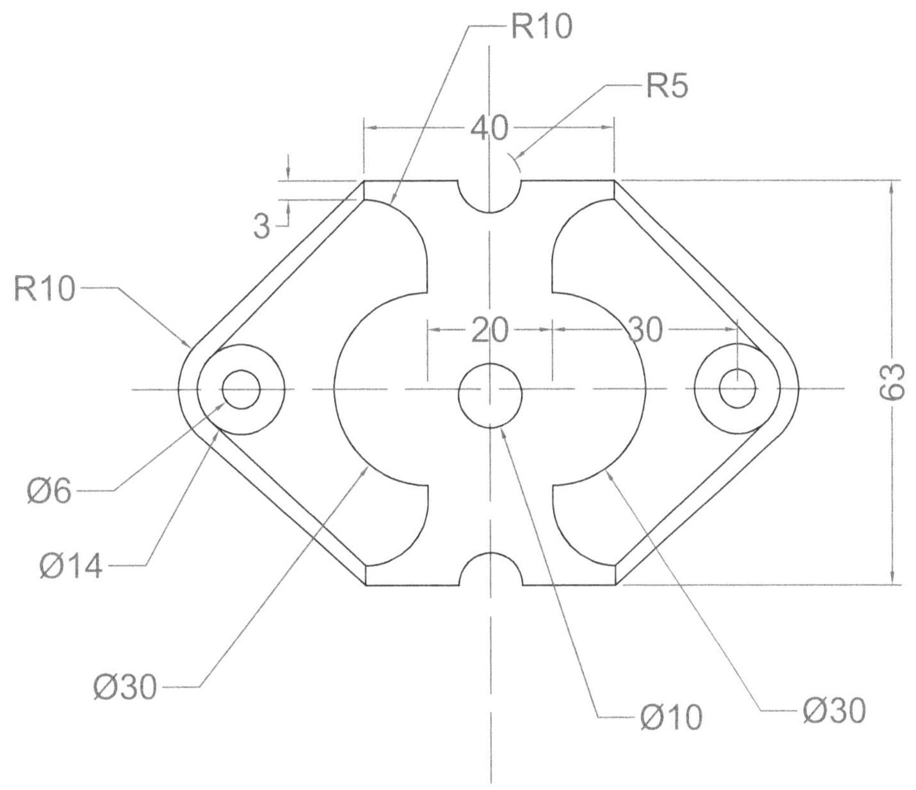

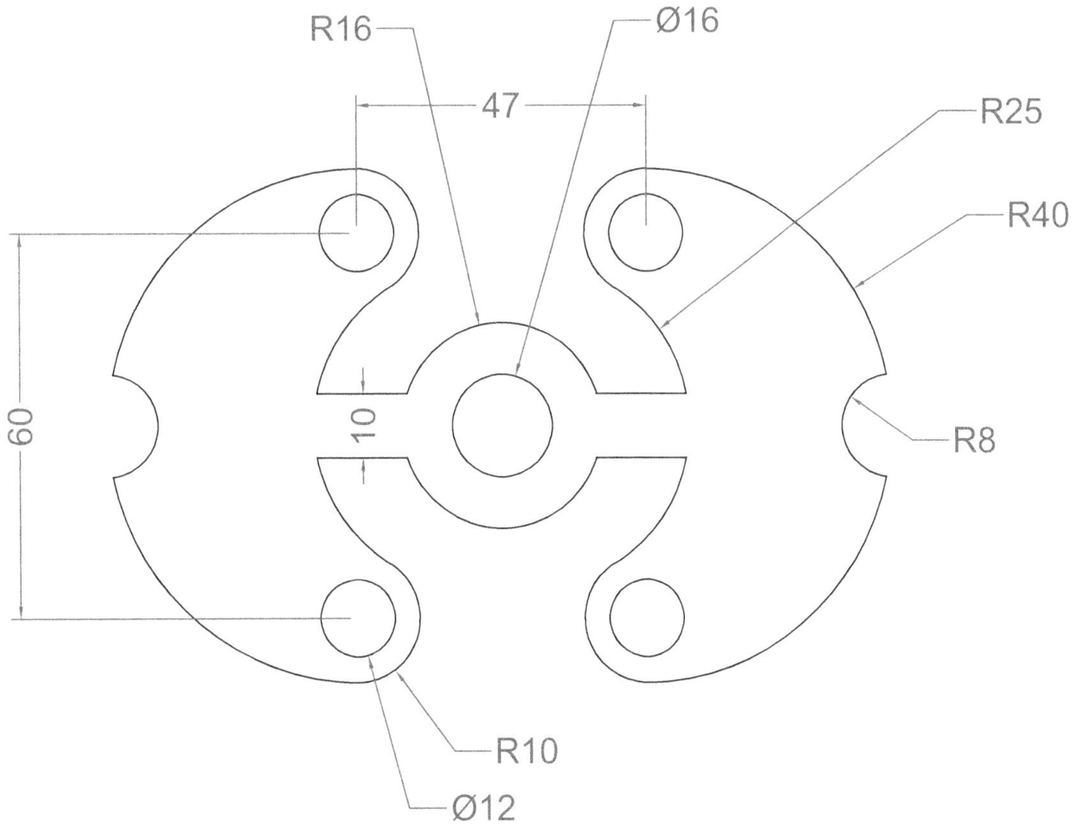

ALL DIMENSIONS ARE IN MM

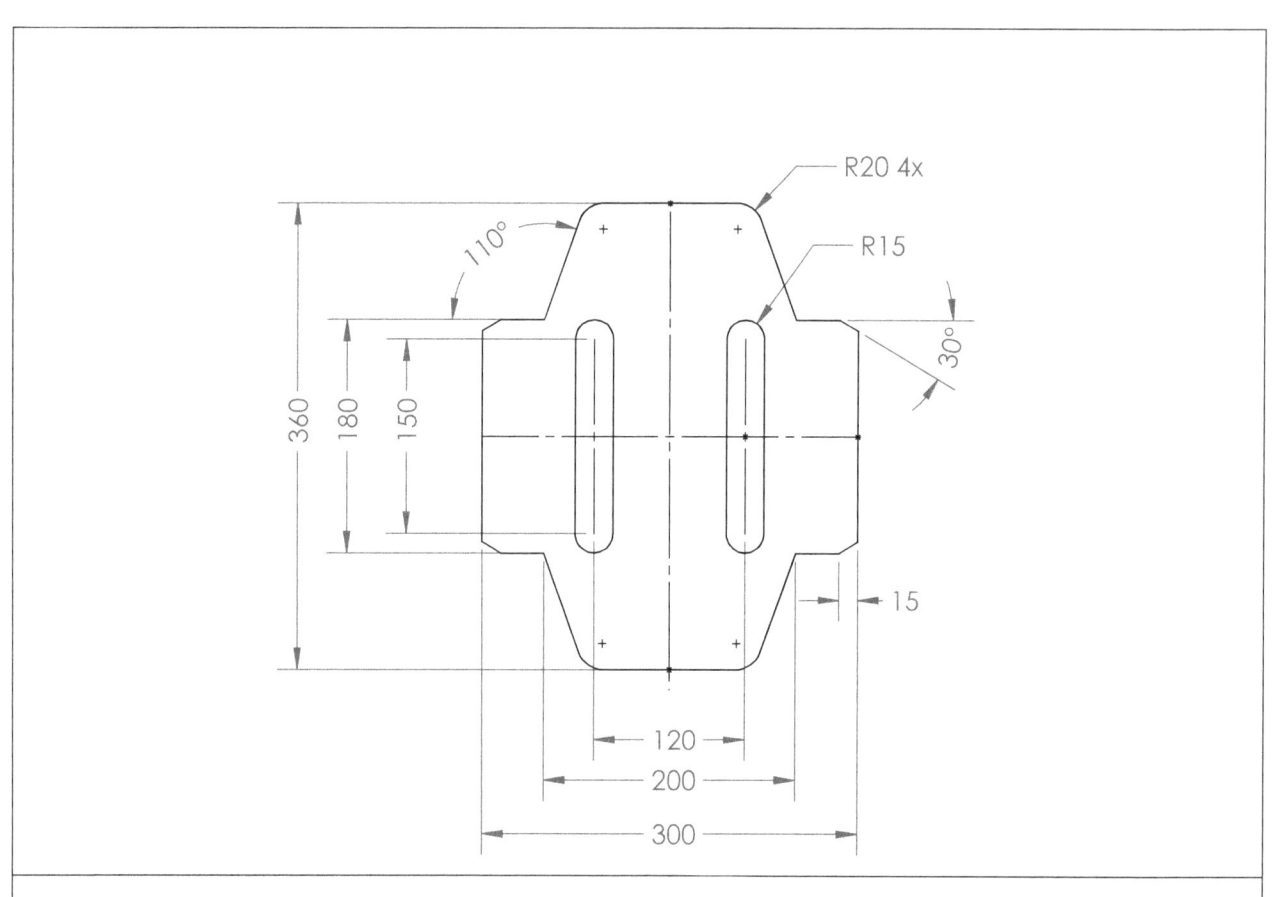
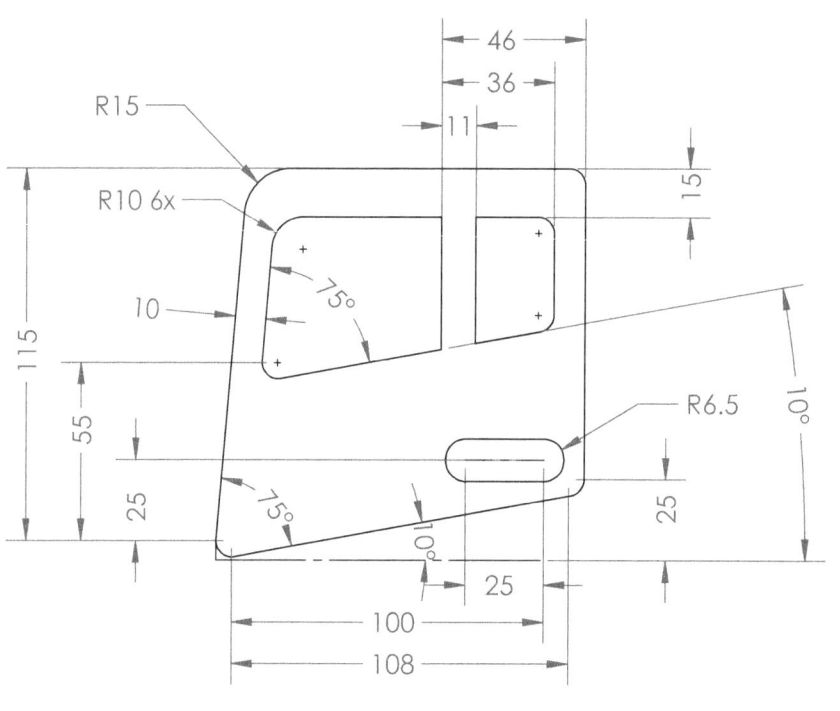

ALL DIMENSIONS ARE IN MM

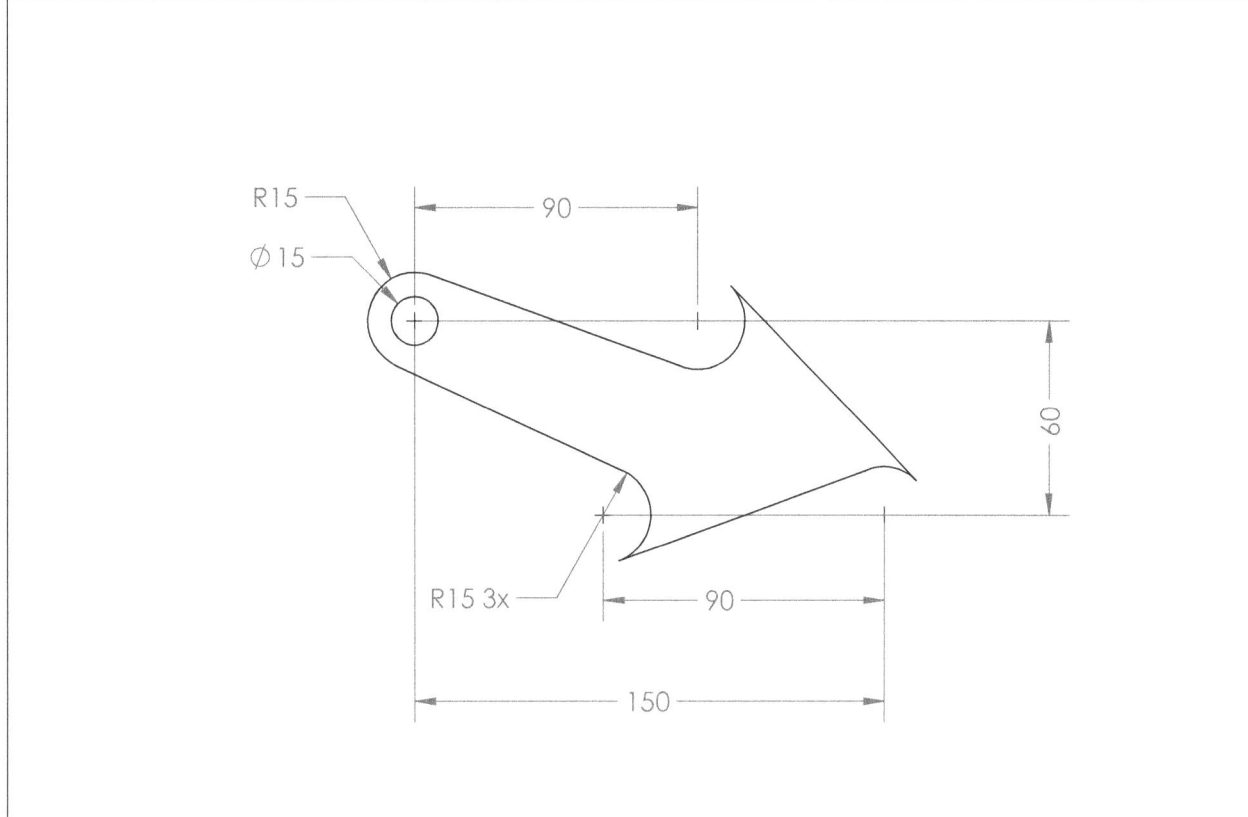
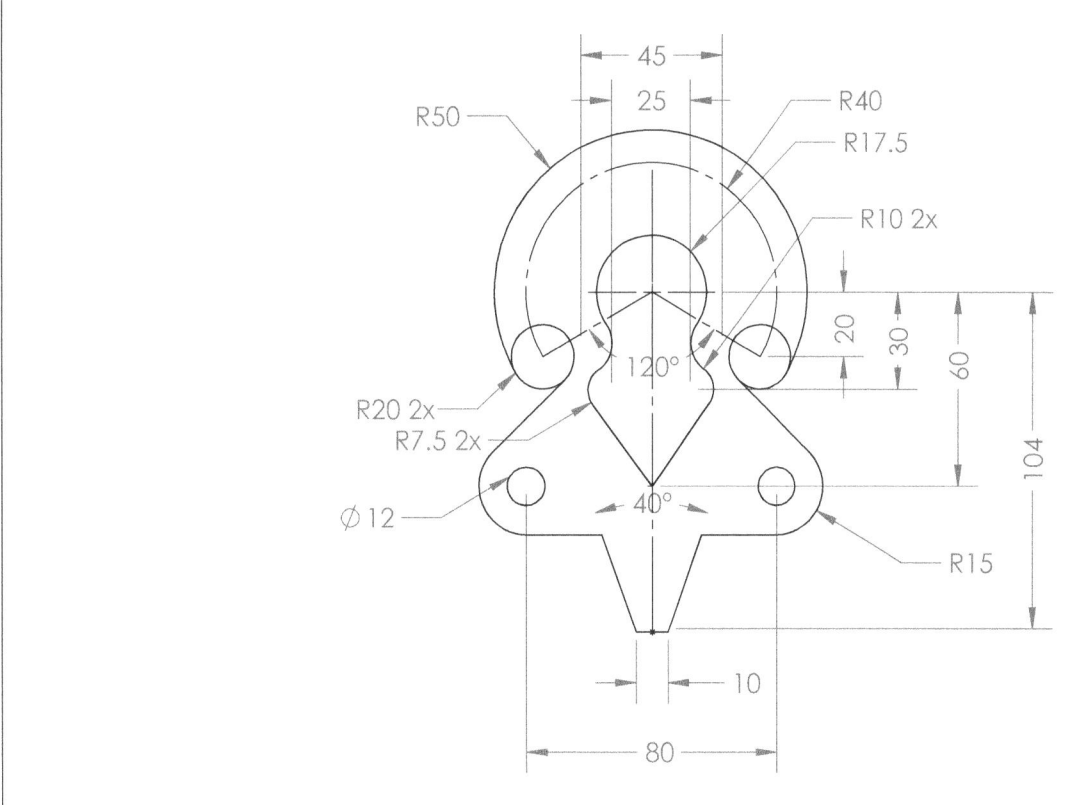

ALL DIMENSIONS ARE IN MM

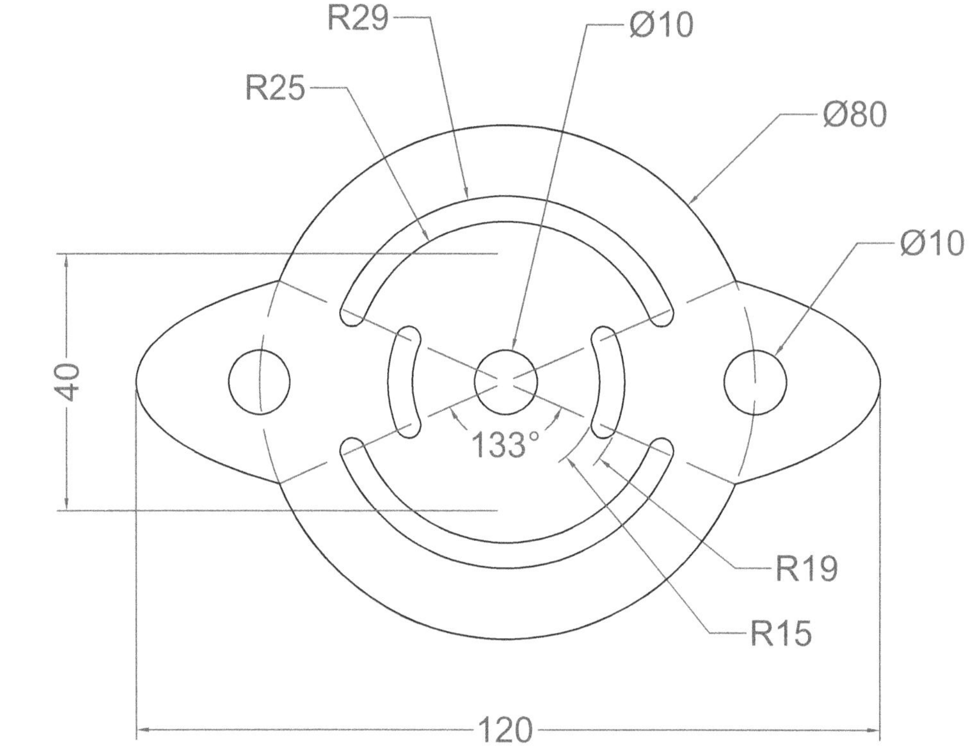

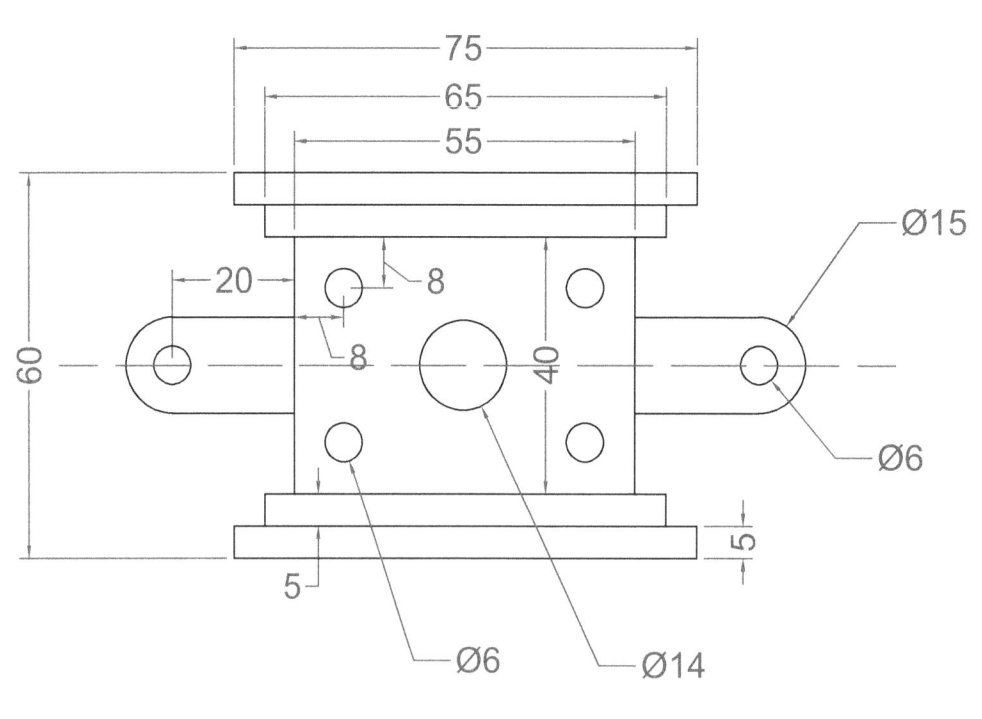

ALL DIMENSIONS ARE IN MM

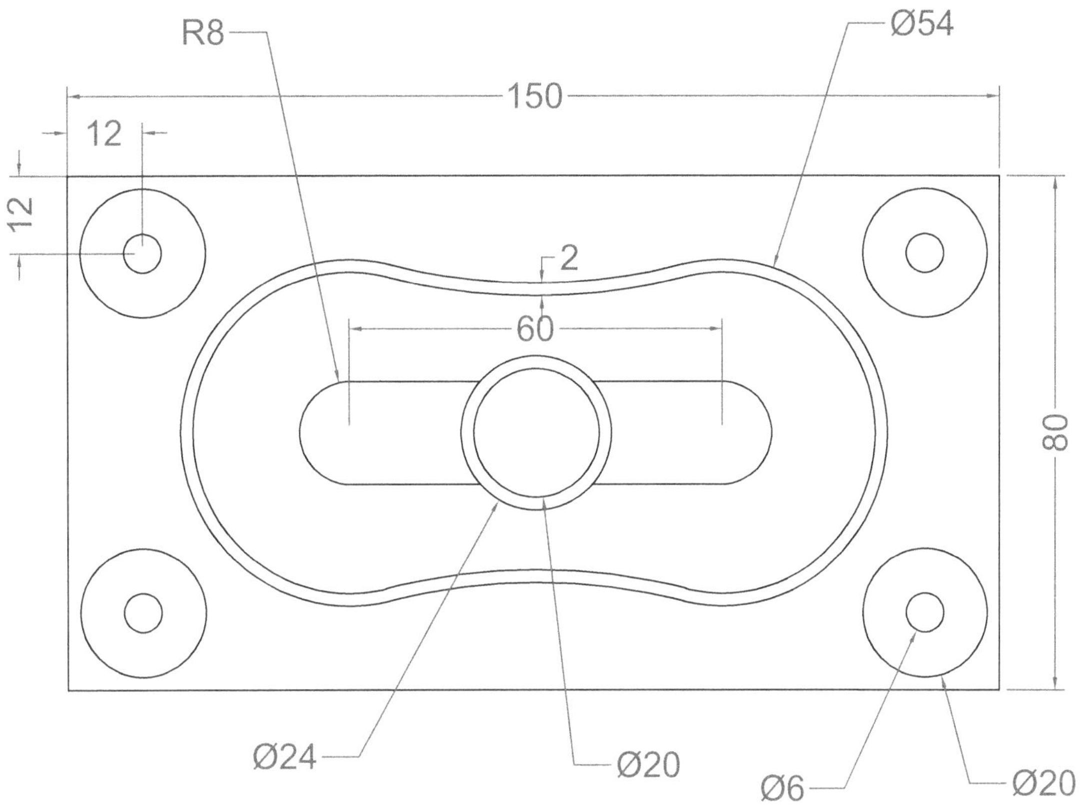

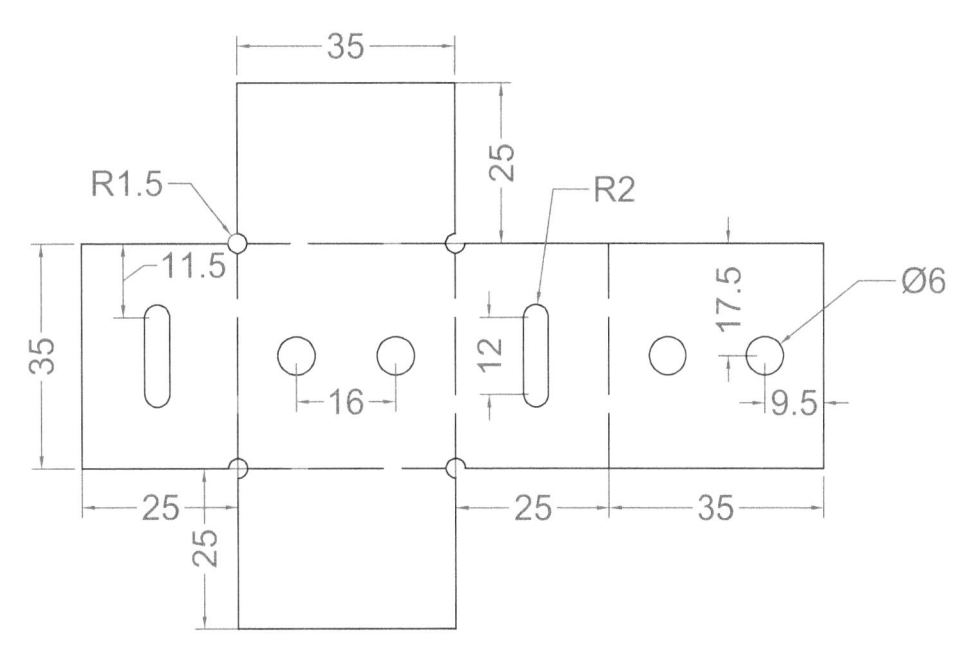

ALL DIMENSIONS ARE IN MM

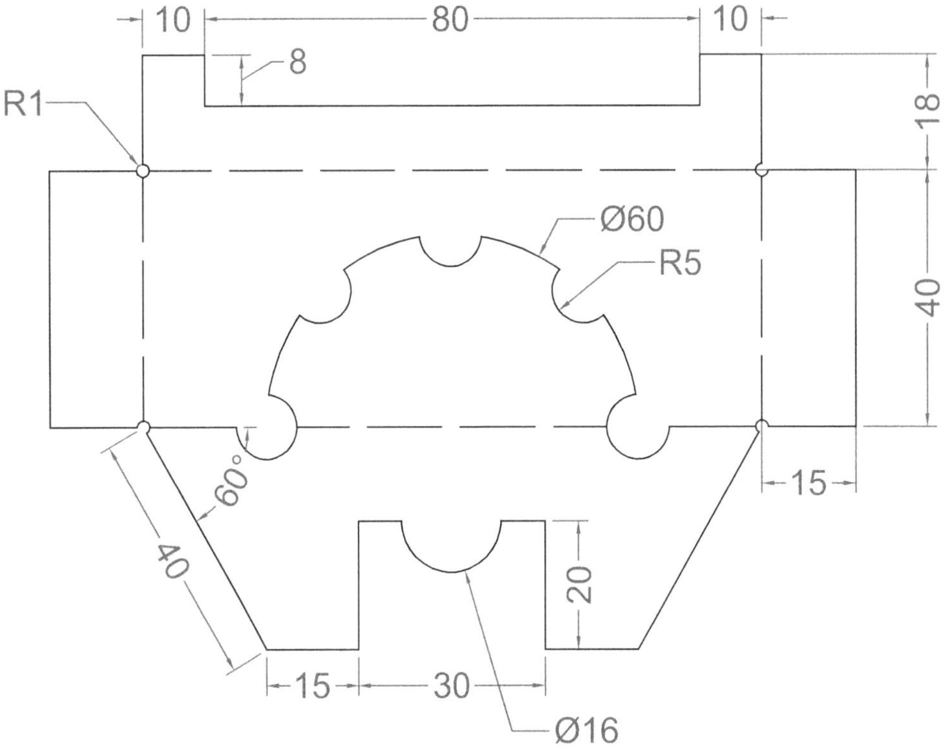

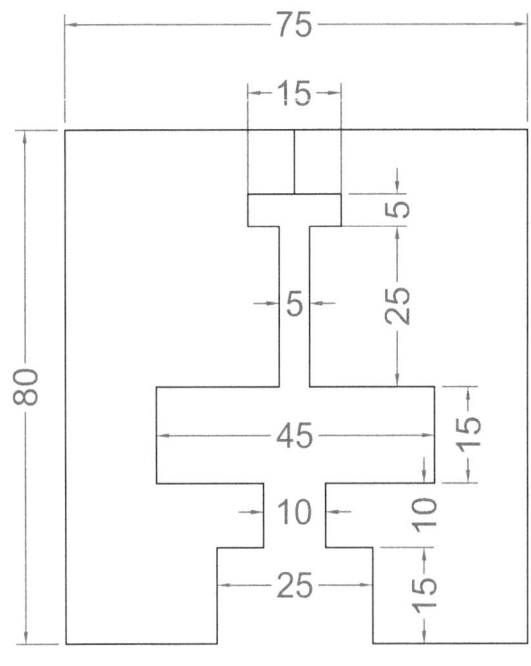

ALL DIMENSIONS ARE IN MM

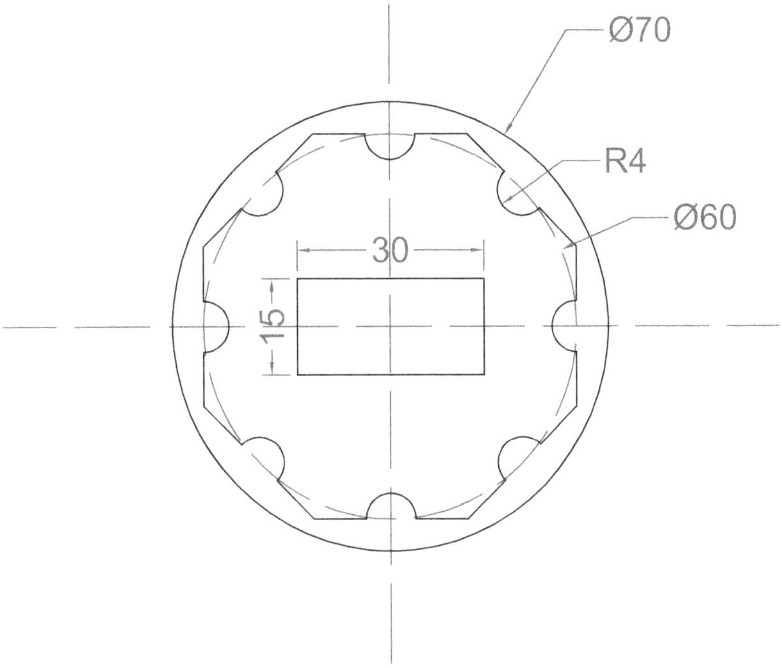

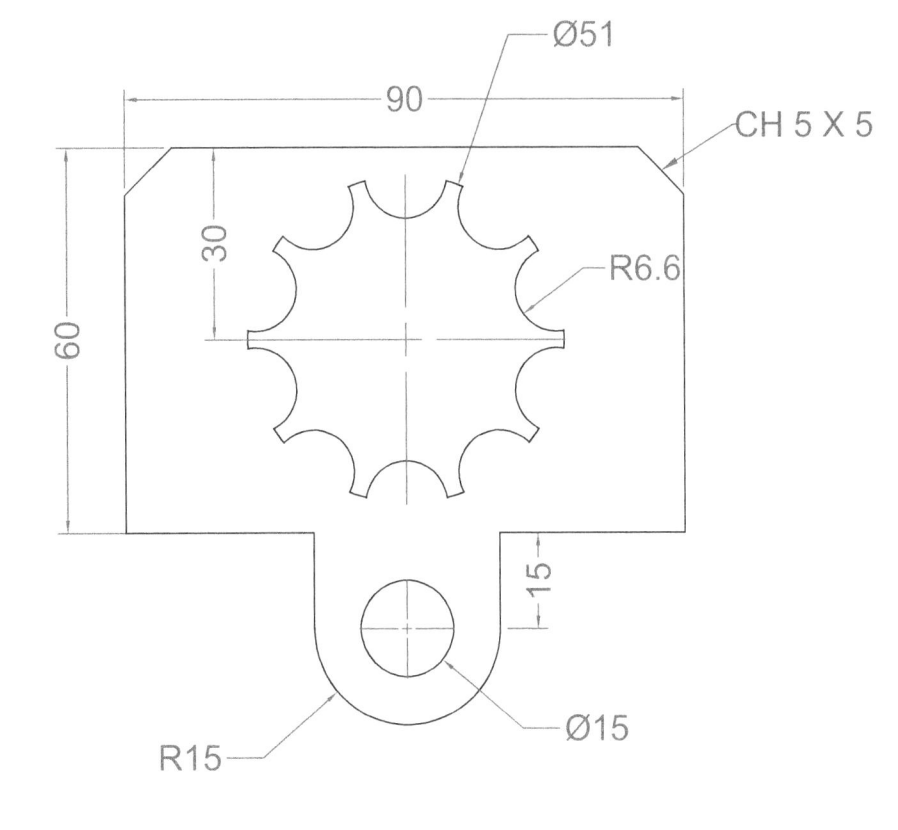

ALL DIMENSIONS ARE IN MM

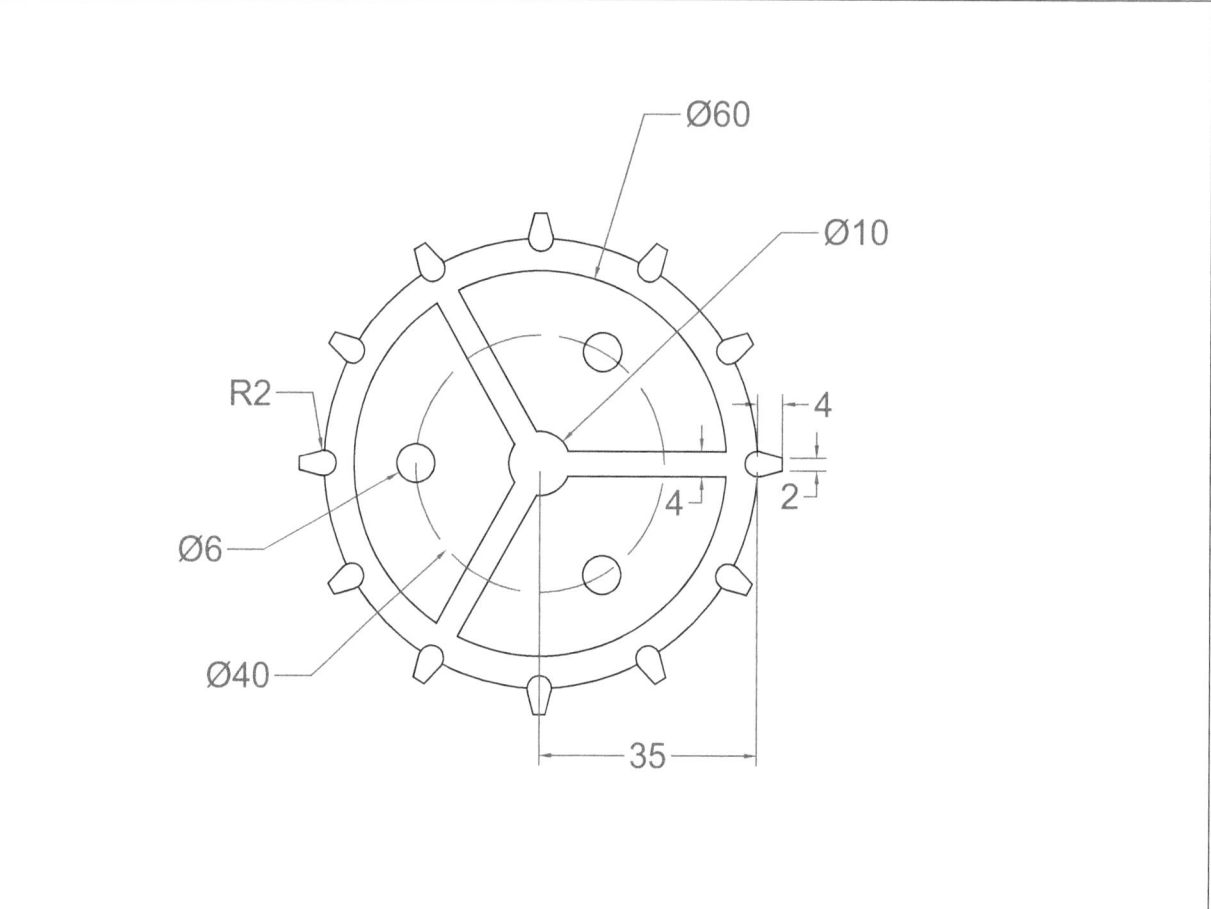
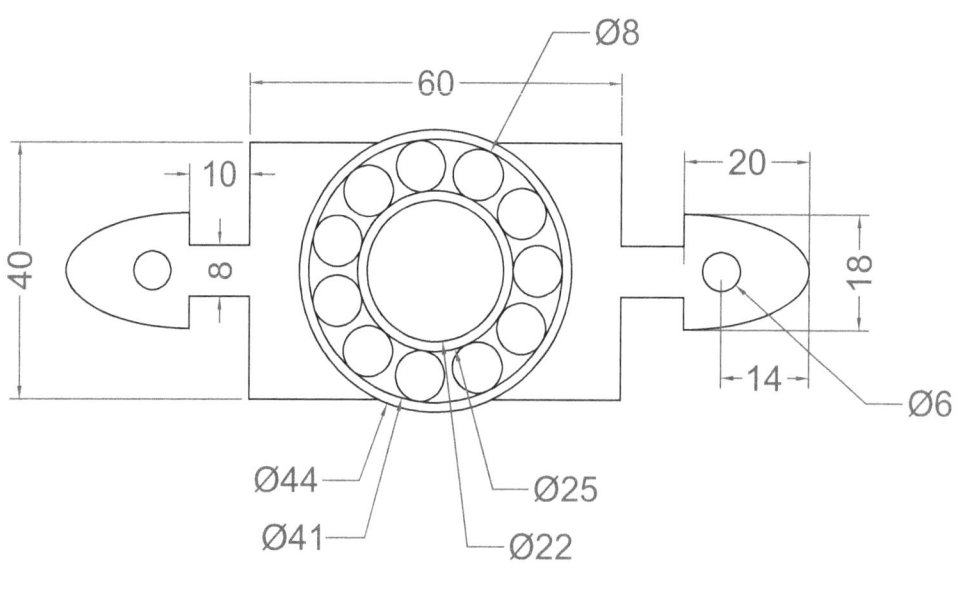

ALL DIMENSIONS ARE IN MM

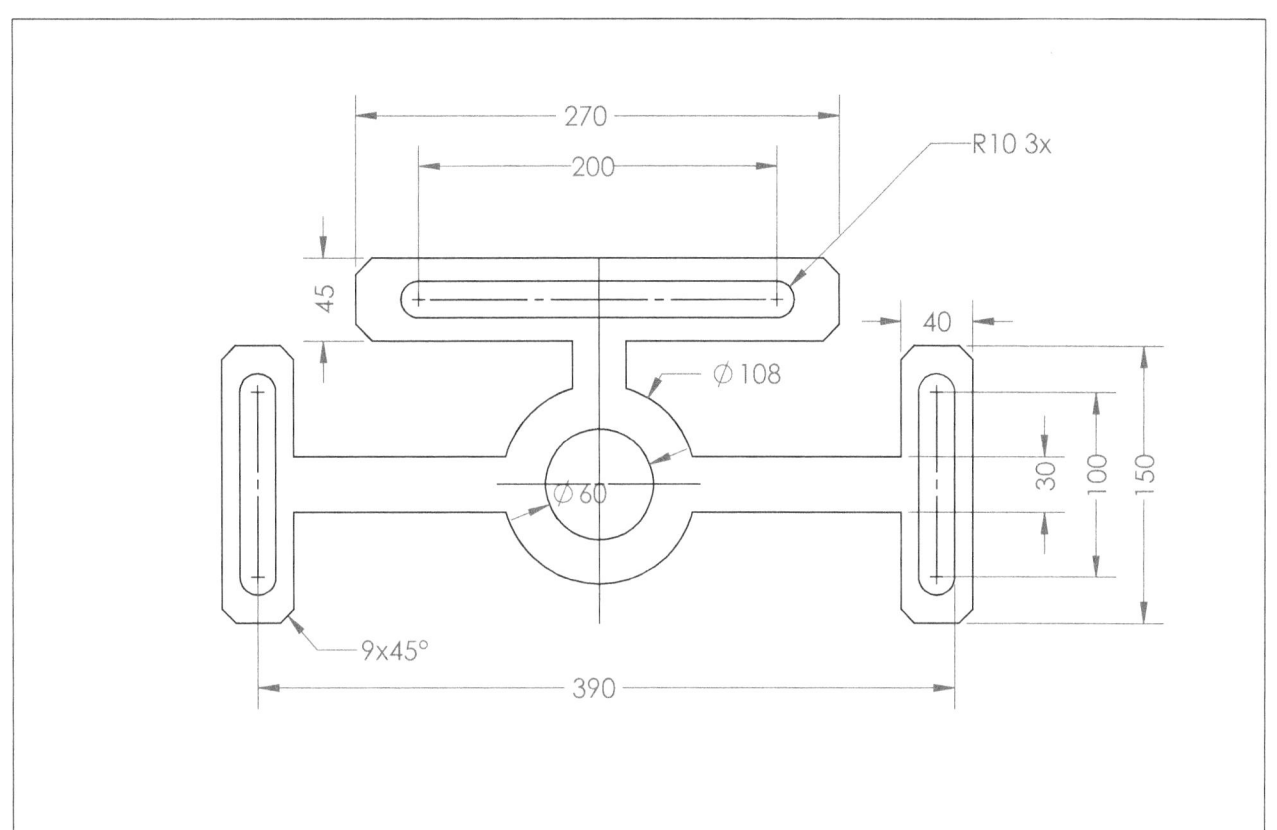
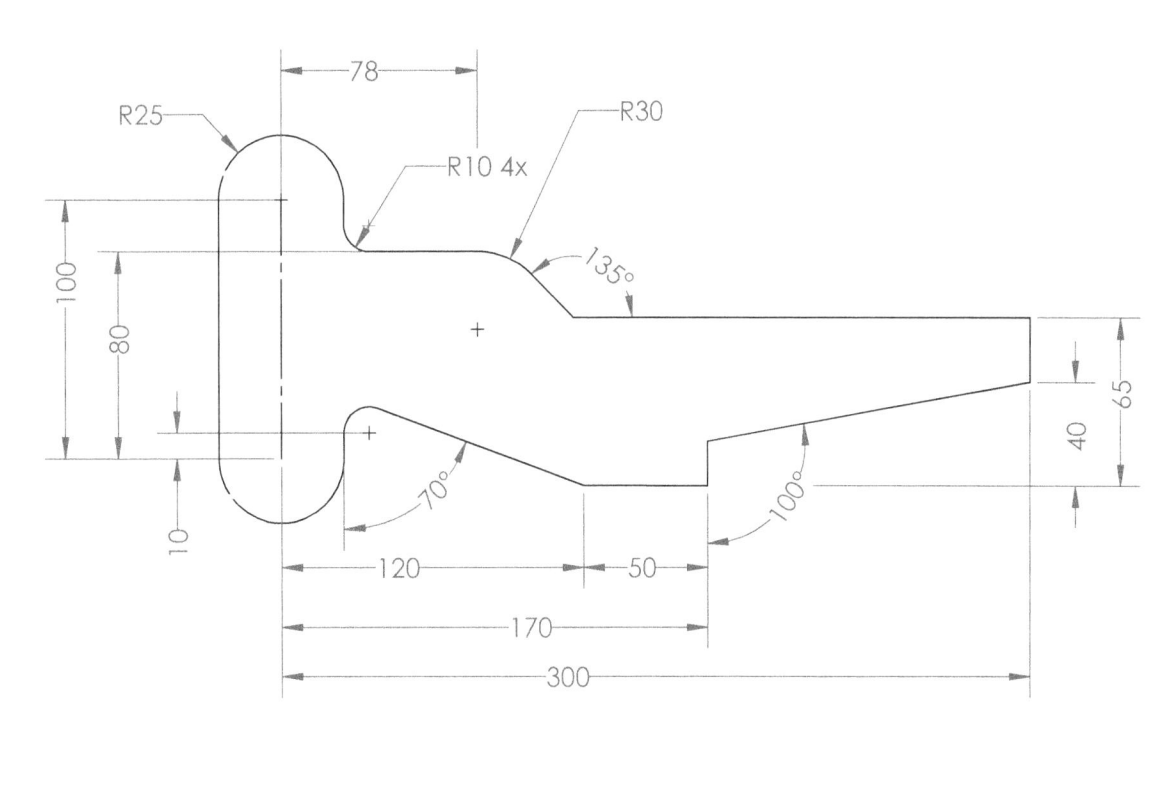

ALL DIMENSIONS ARE IN MM

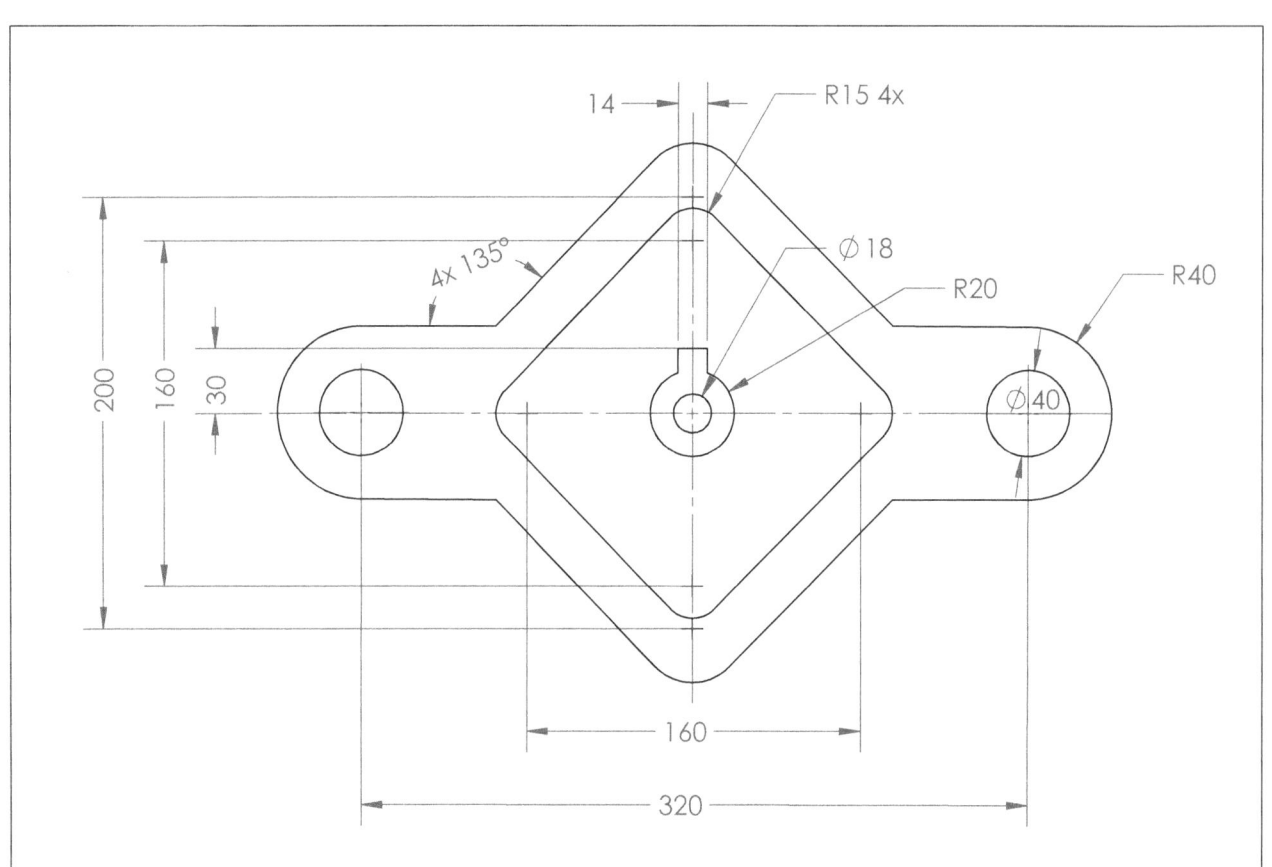
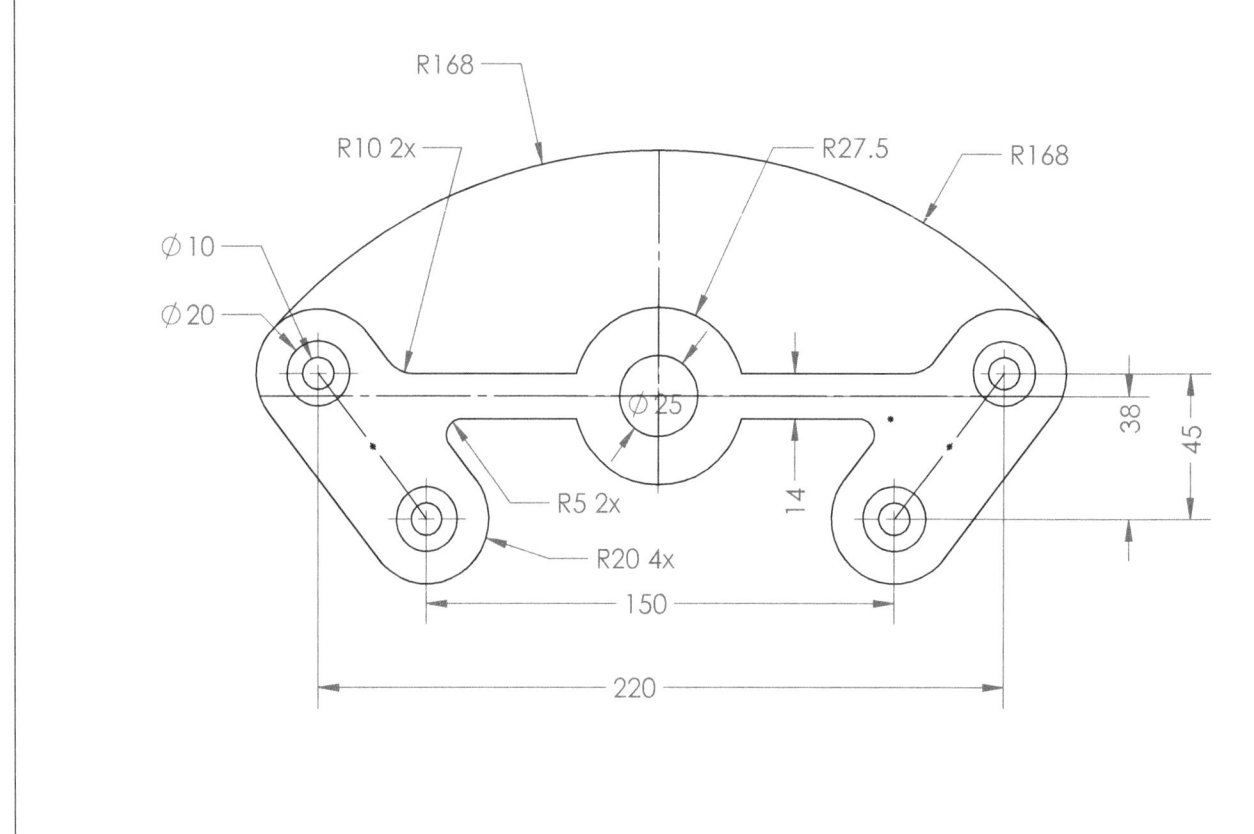

ALL DIMENSIONS ARE IN MM

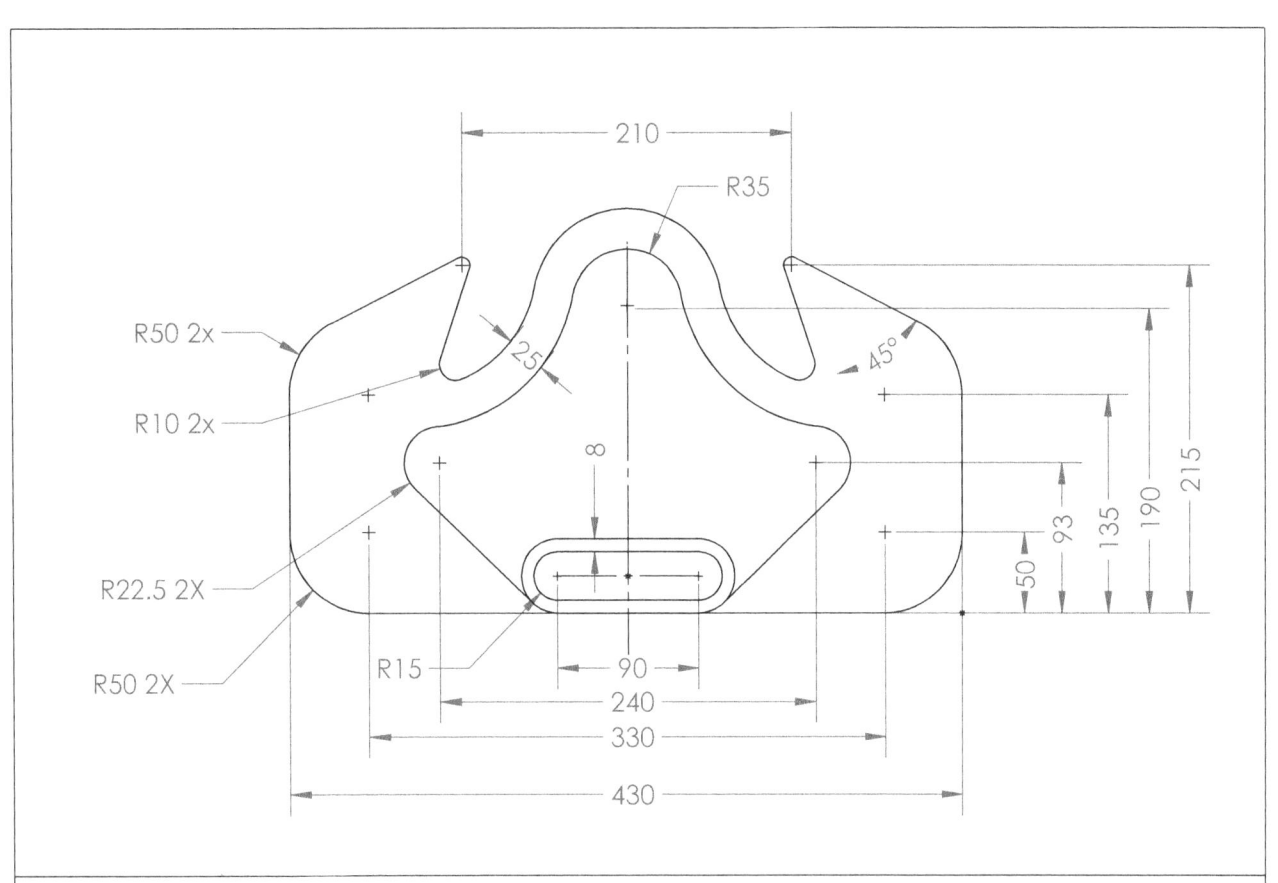

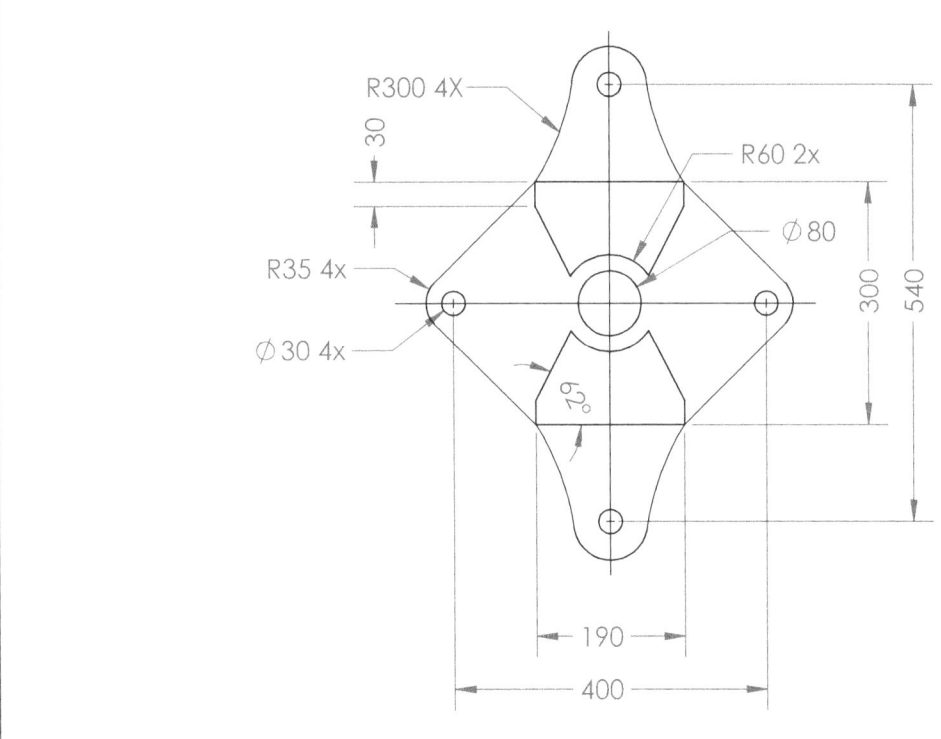

ALL DIMENSIONS ARE IN MM

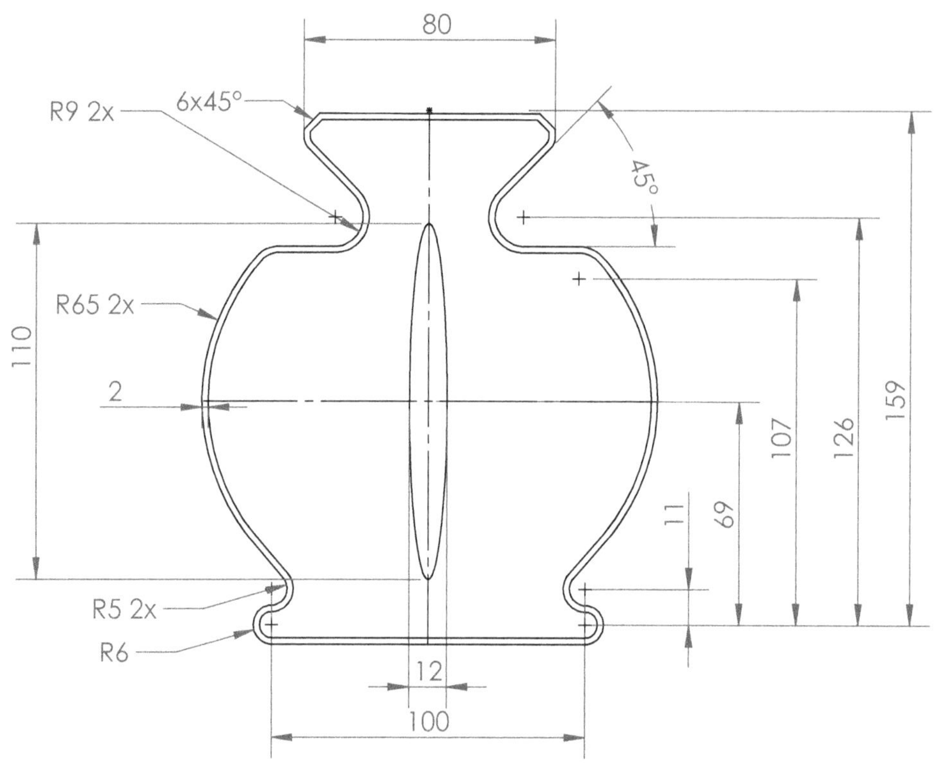

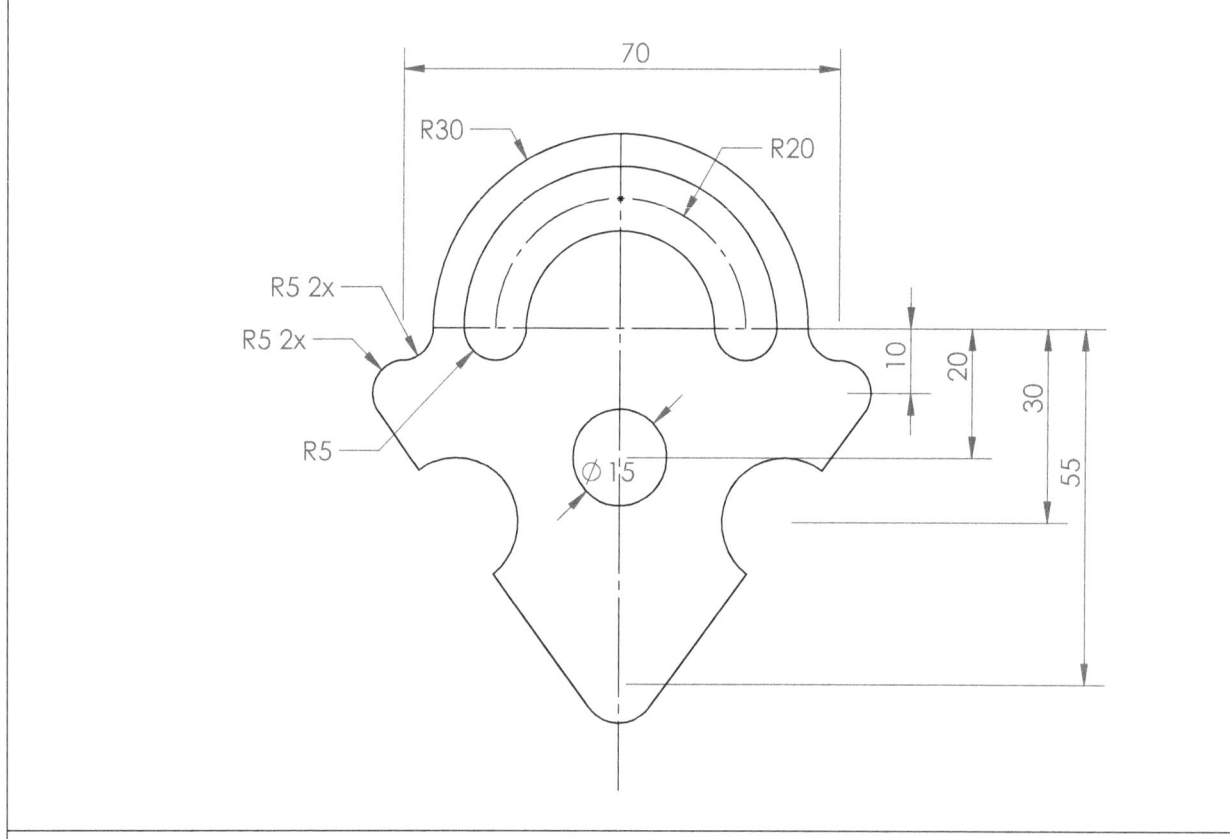

ALL DIMENSIONS ARE IN MM

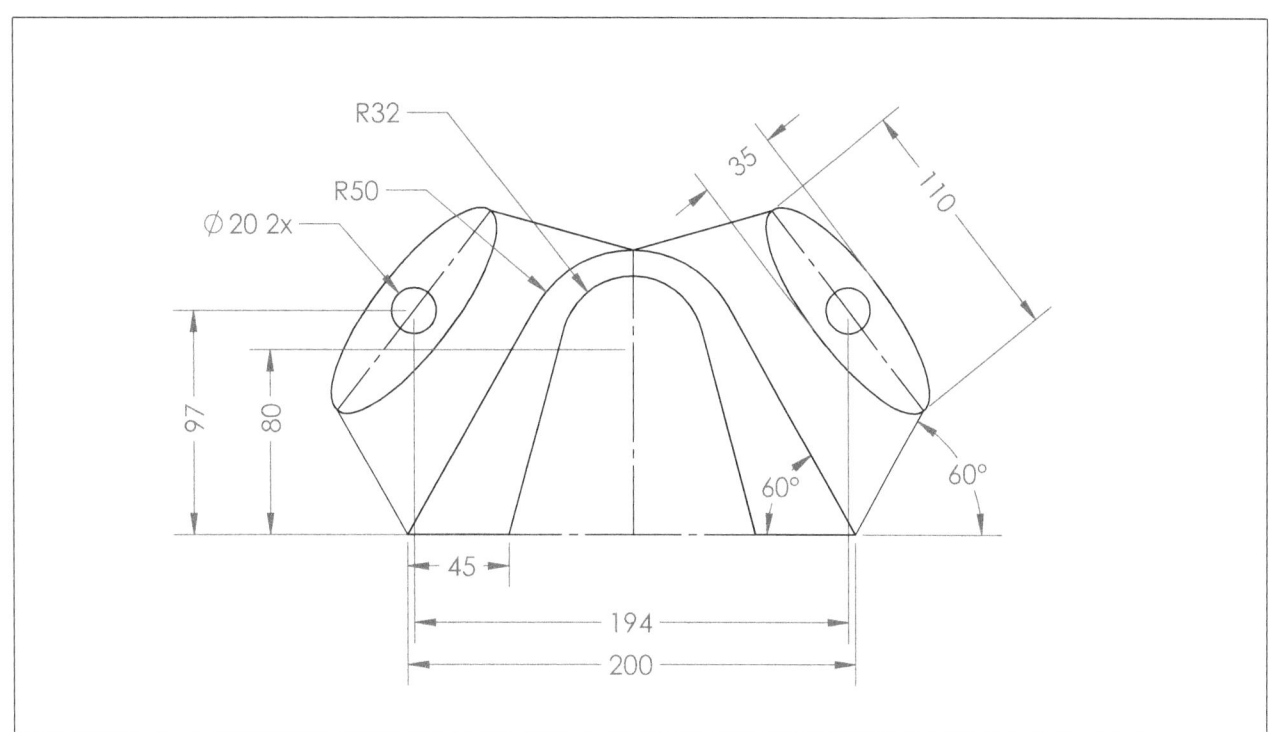
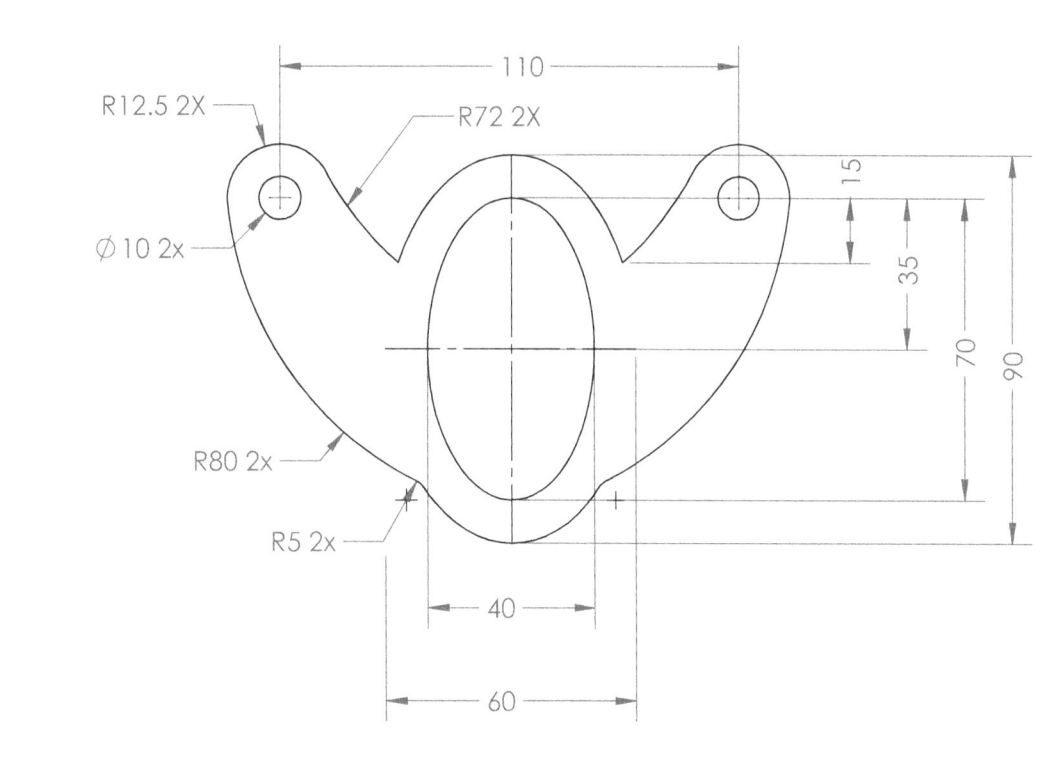

ALL DIMENSIONS ARE IN MM

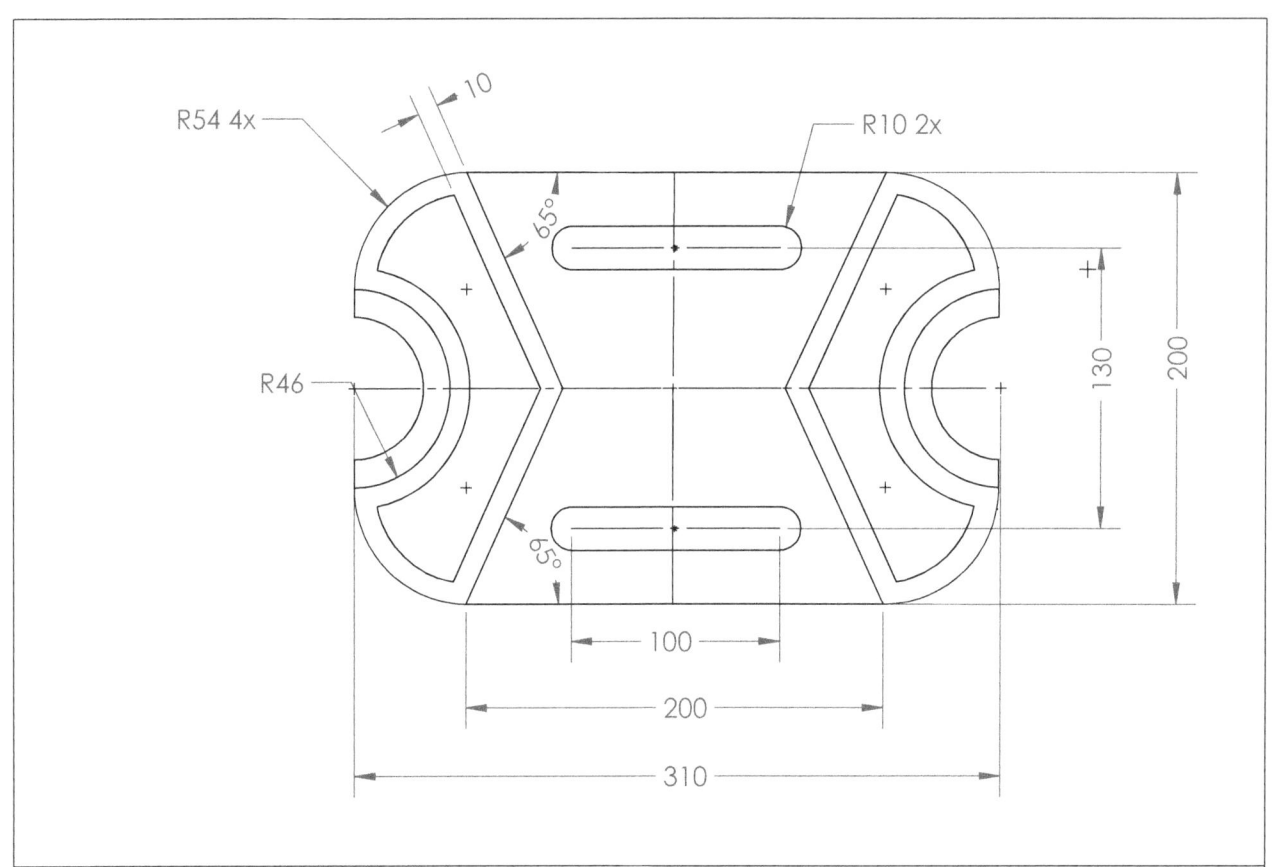
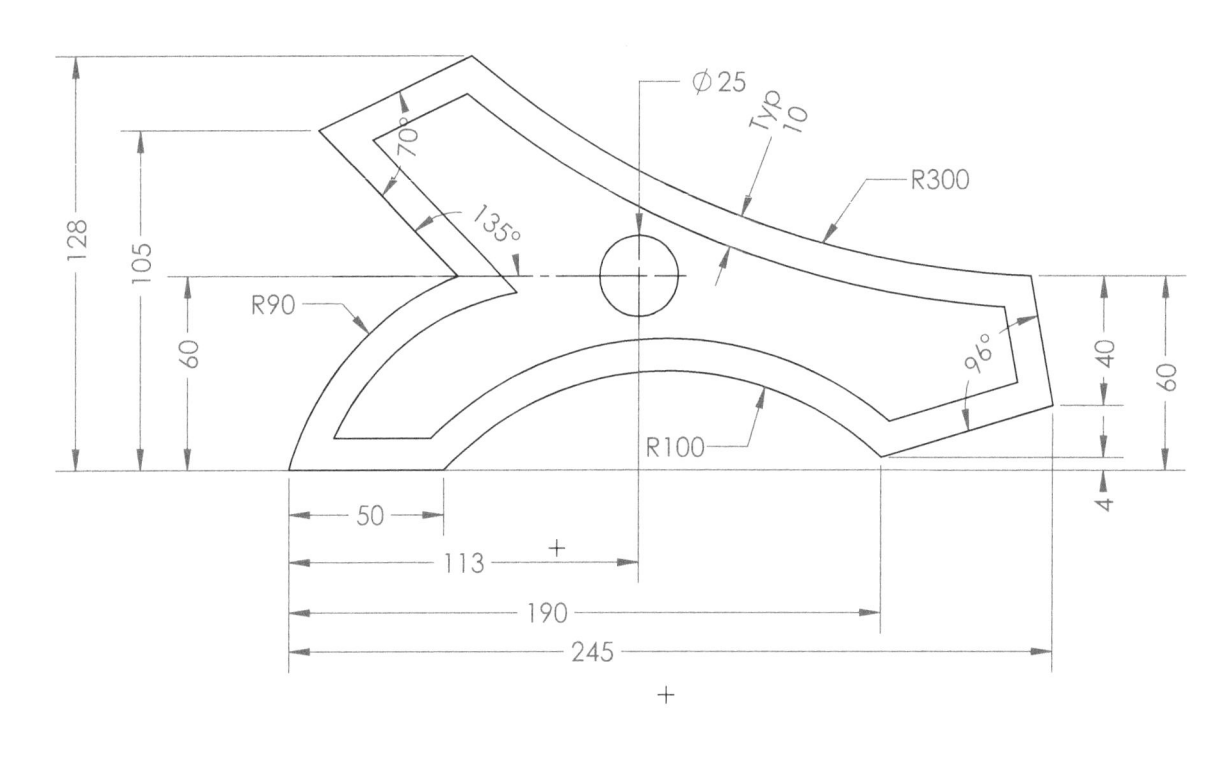

ALL DIMENSIONS ARE IN MM

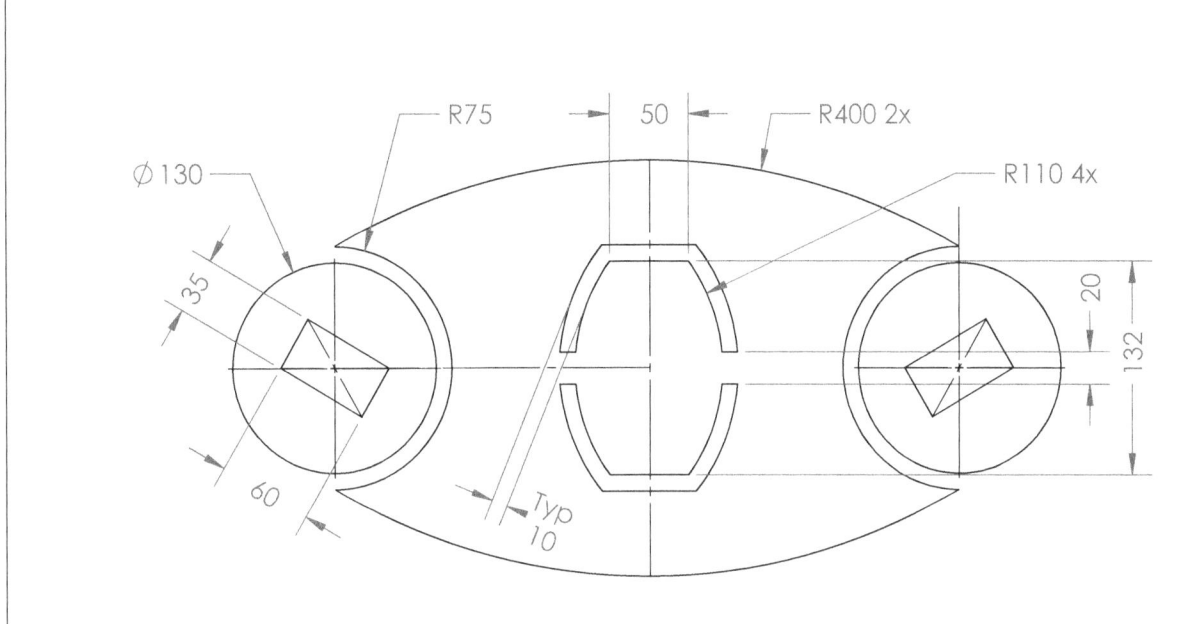

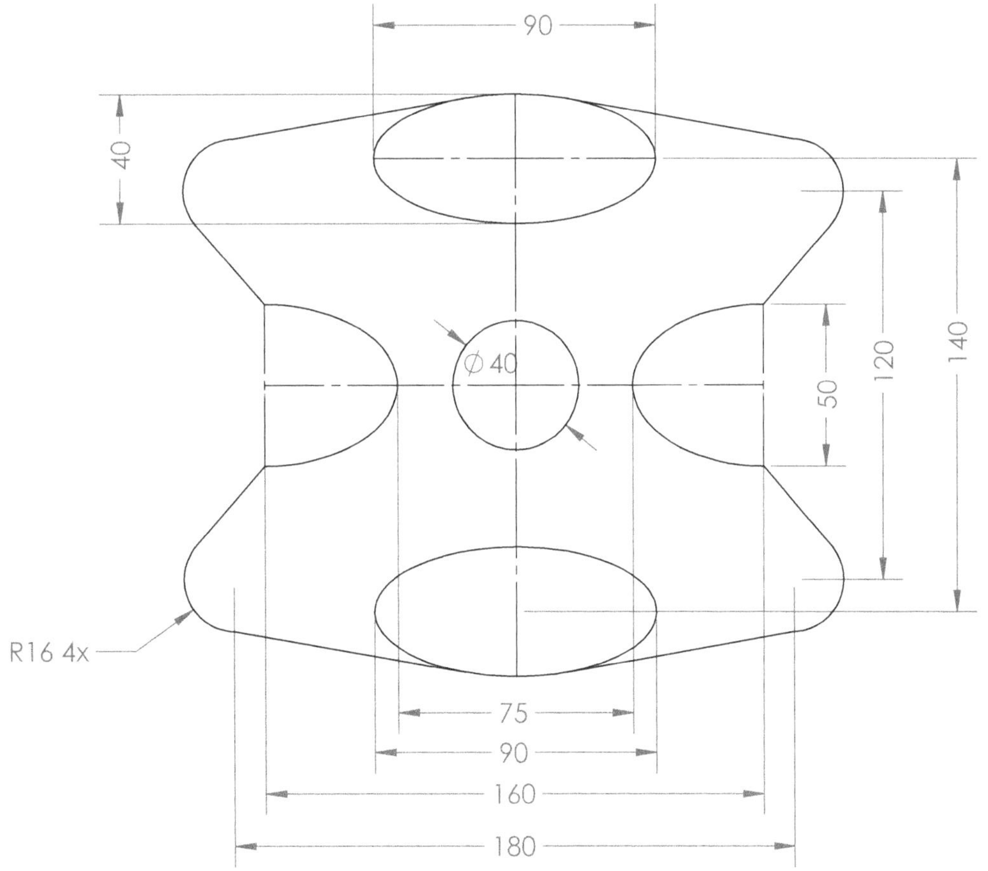

ALL DIMENSIONS ARE IN MM

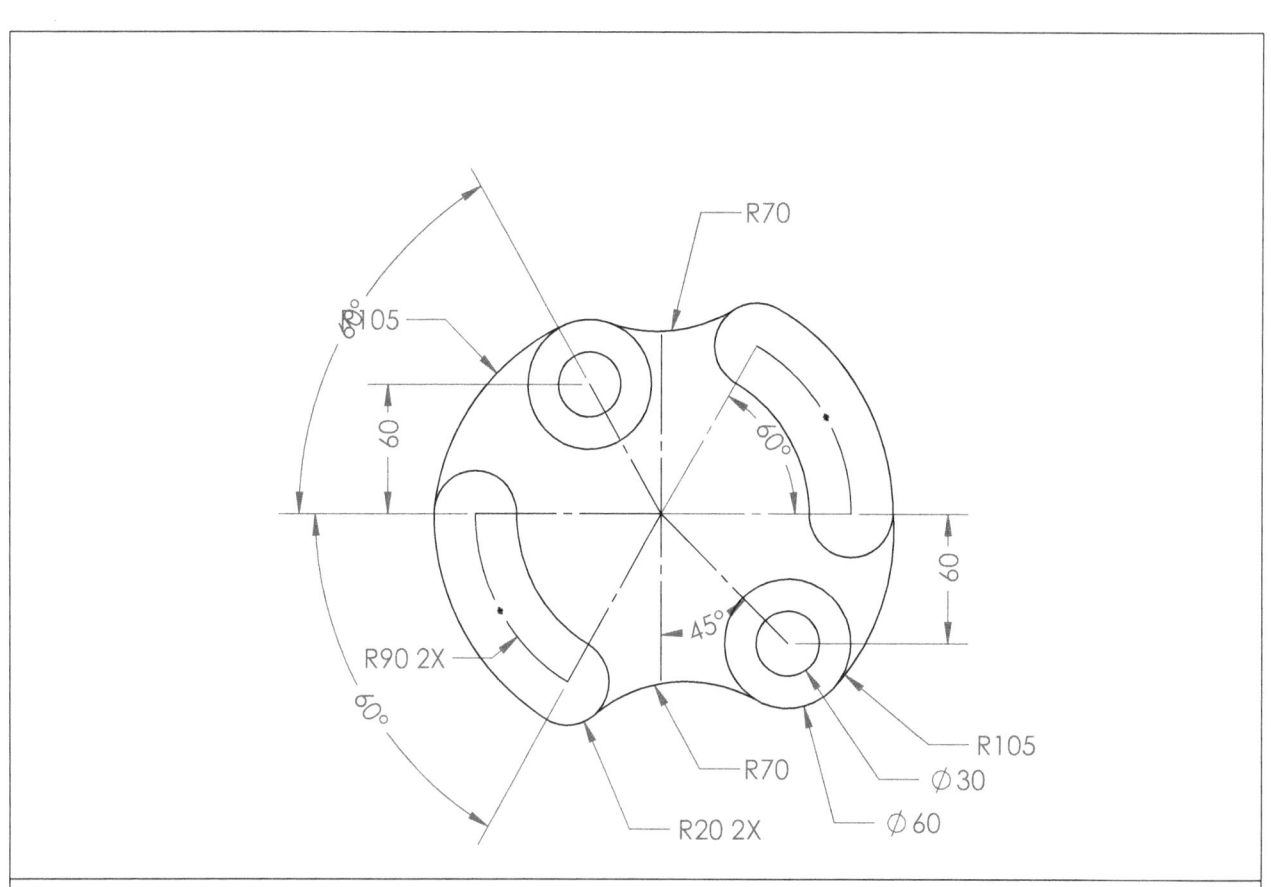
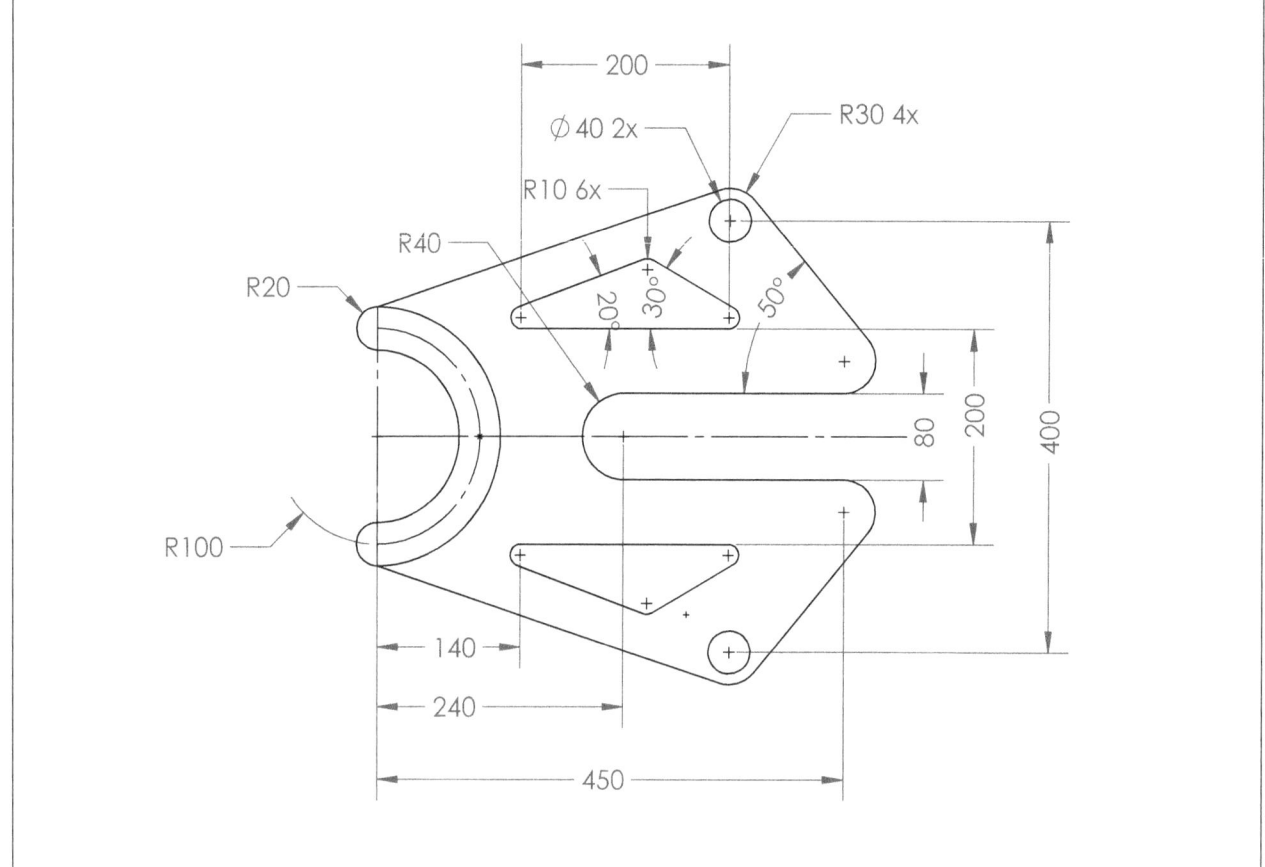

ALL DIMENSIONS ARE IN MM

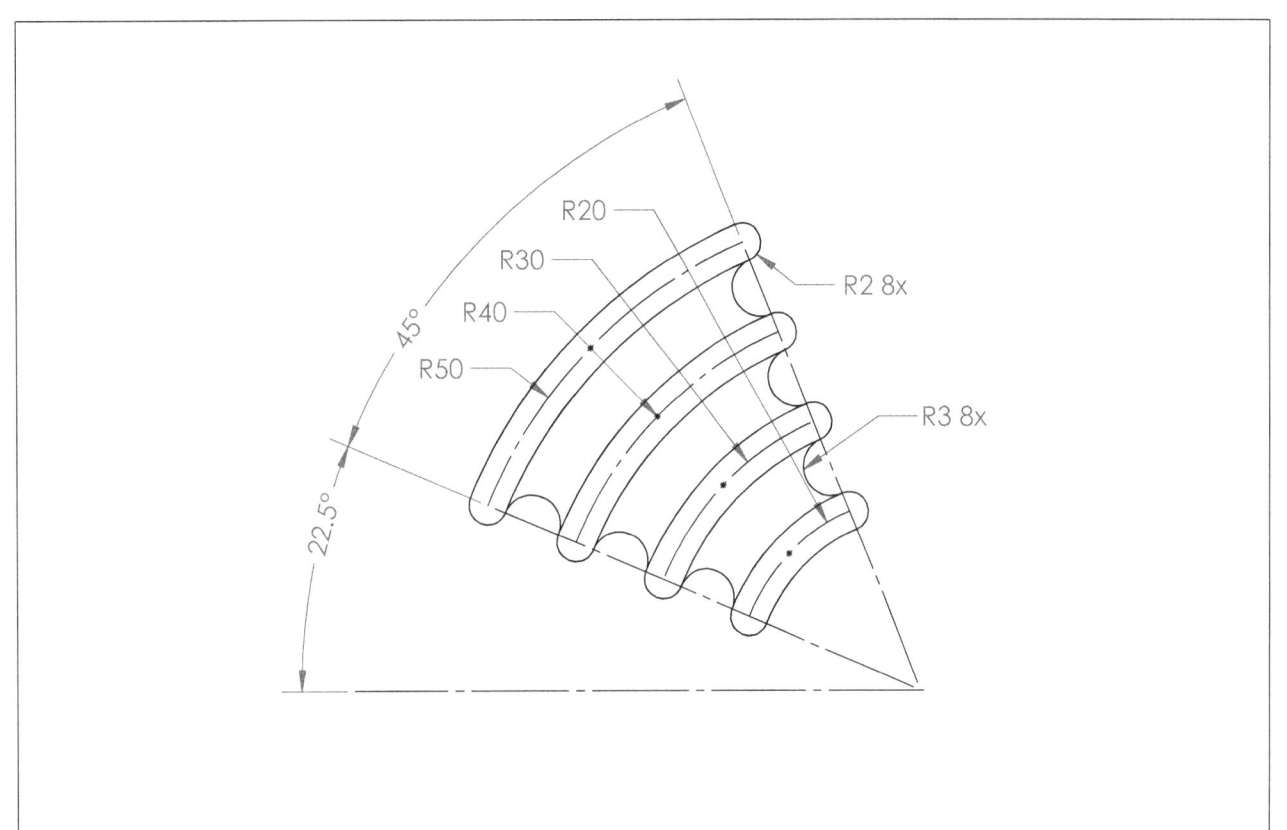

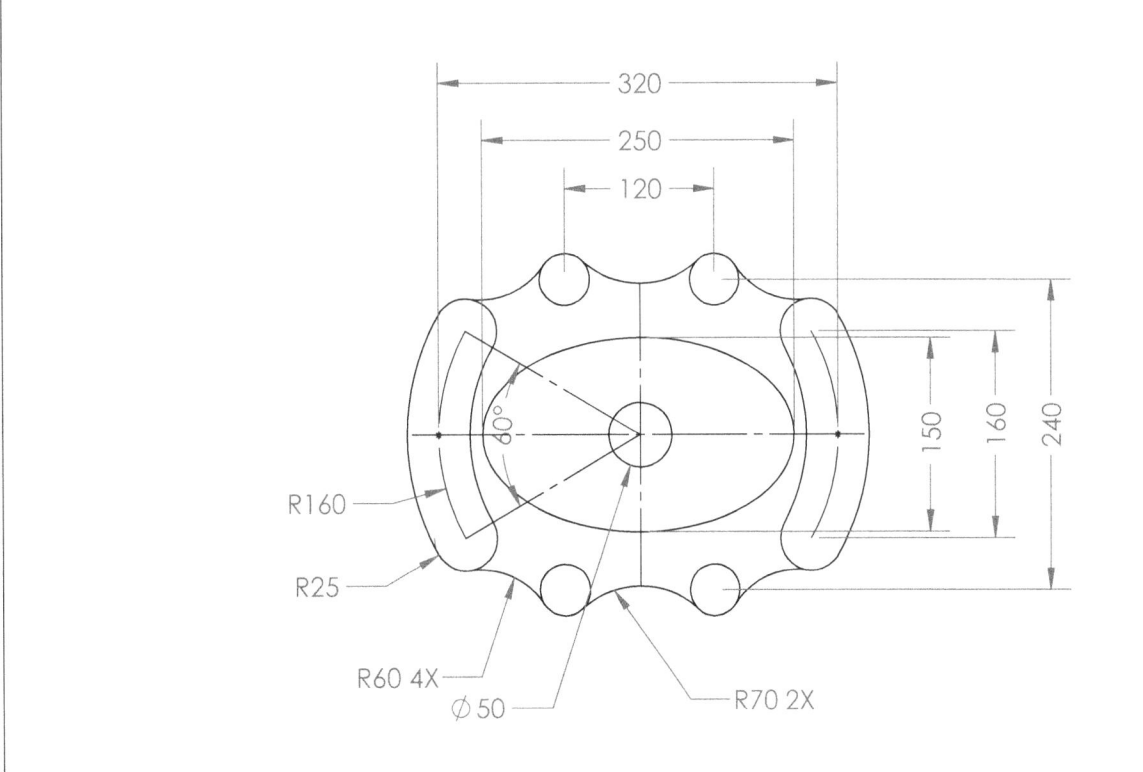

ALL DIMENSIONS ARE IN MM

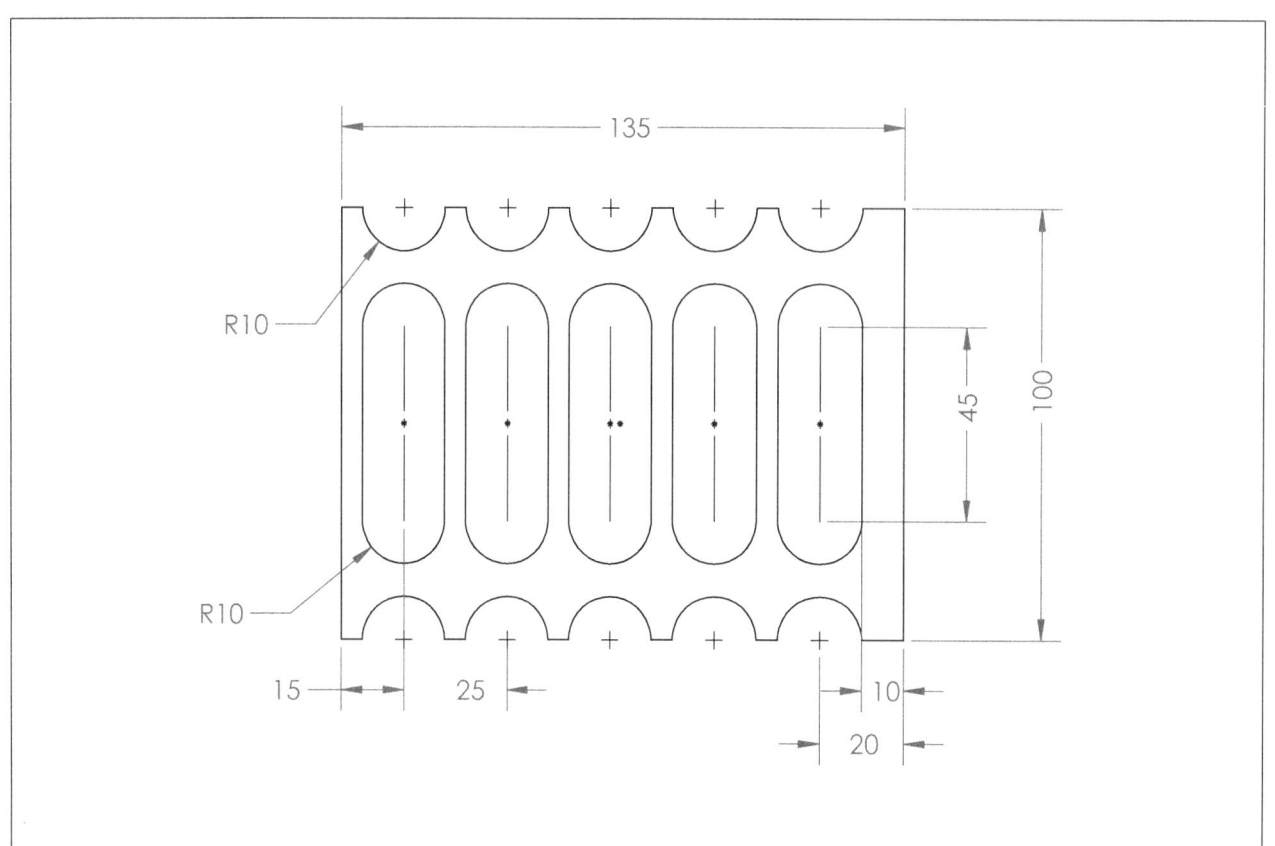

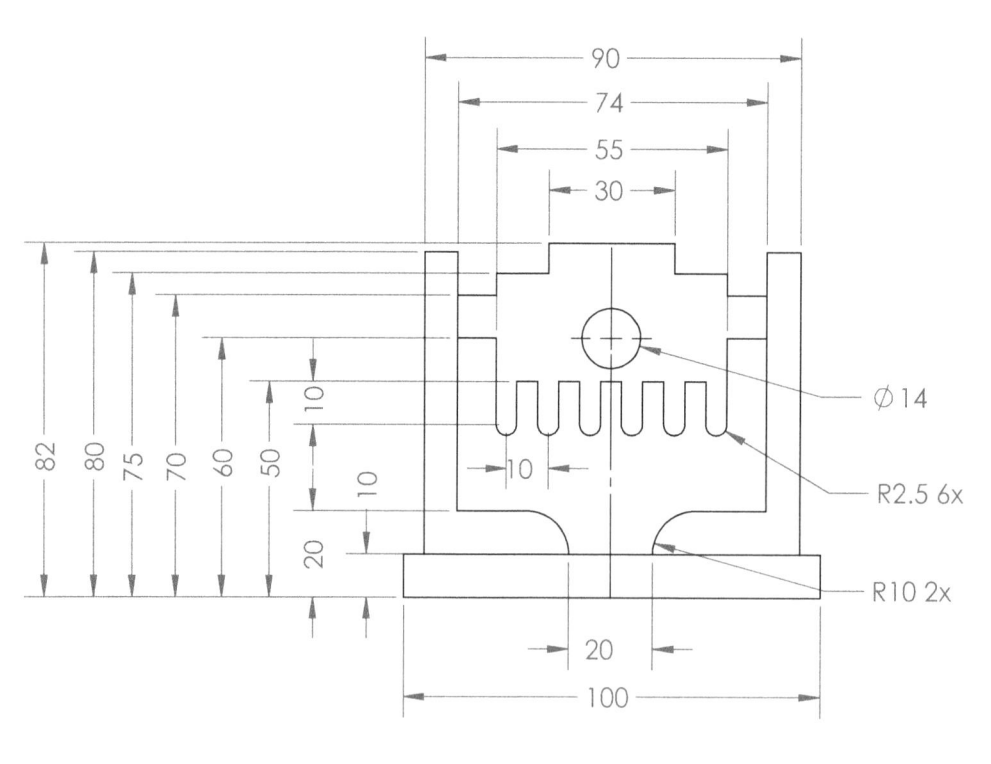

ALL DIMENSIONS ARE IN MM

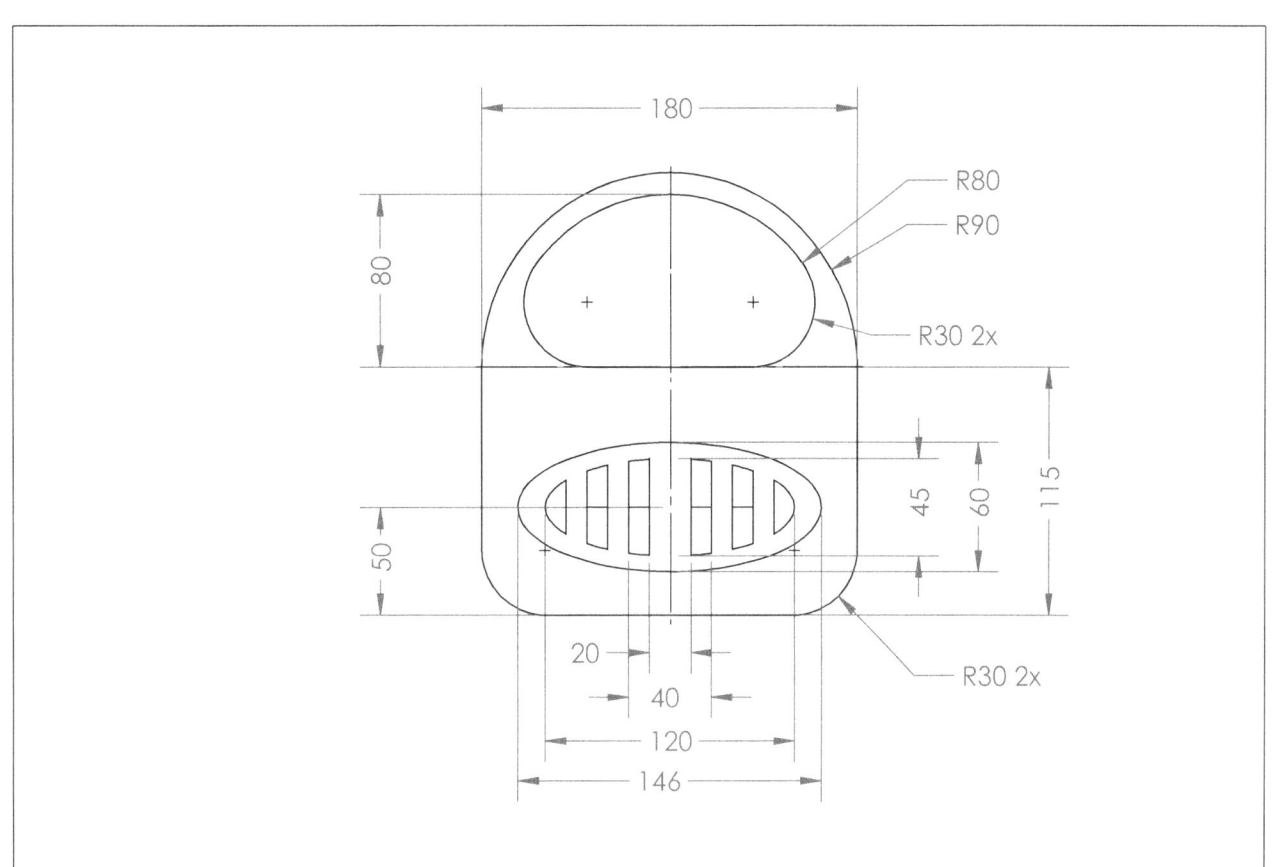
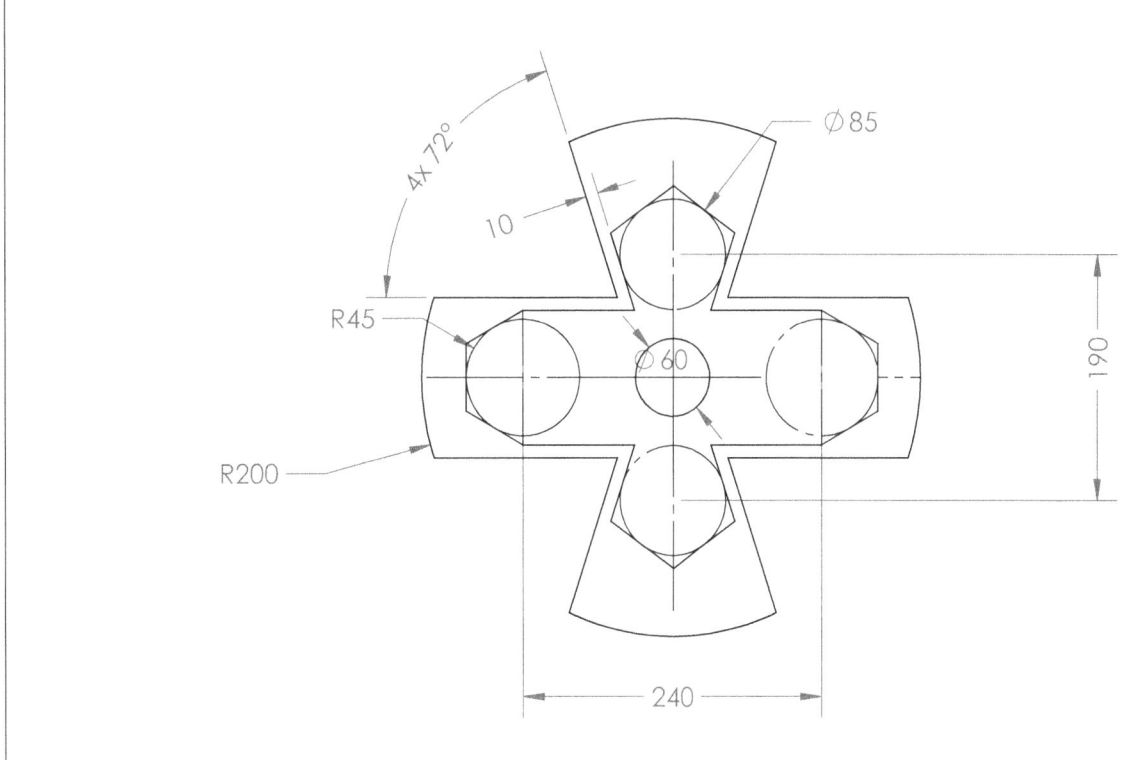

ALL DIMENSIONS ARE IN MM

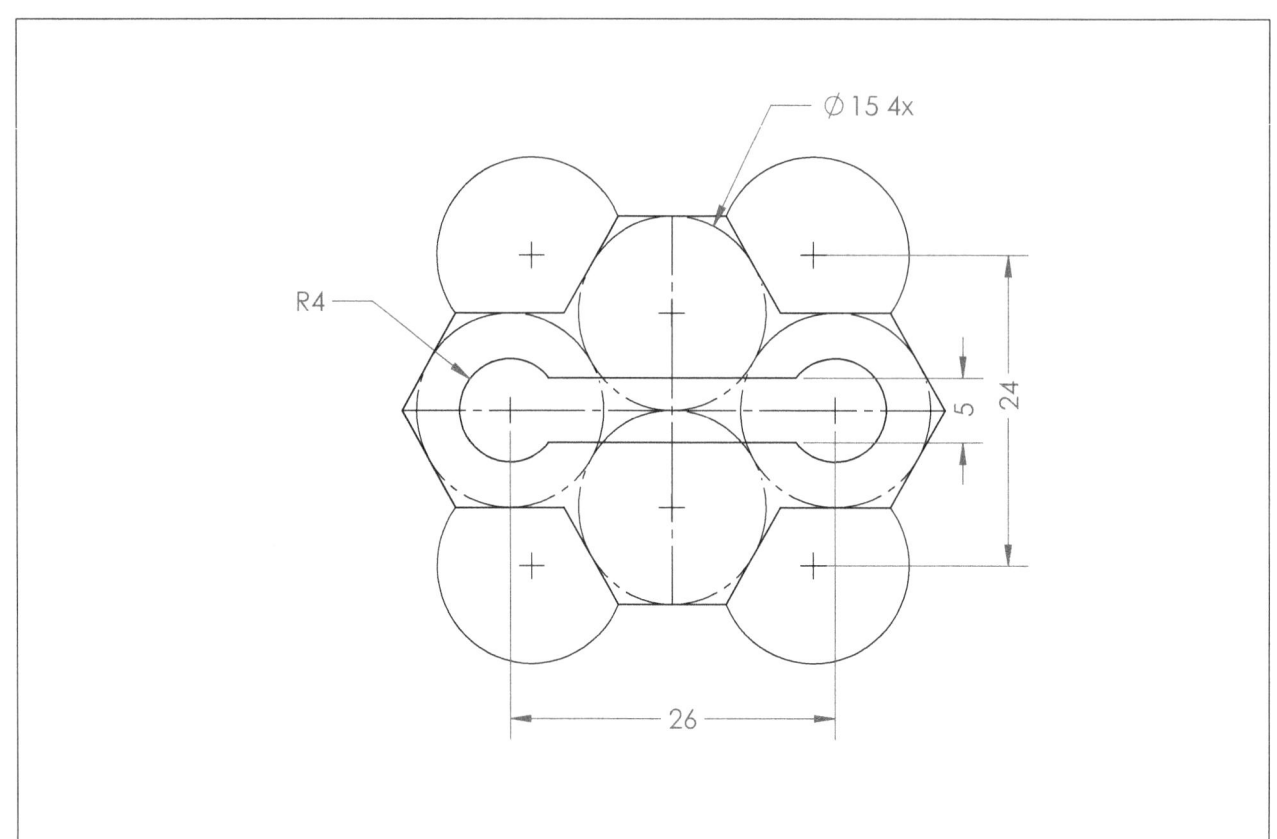
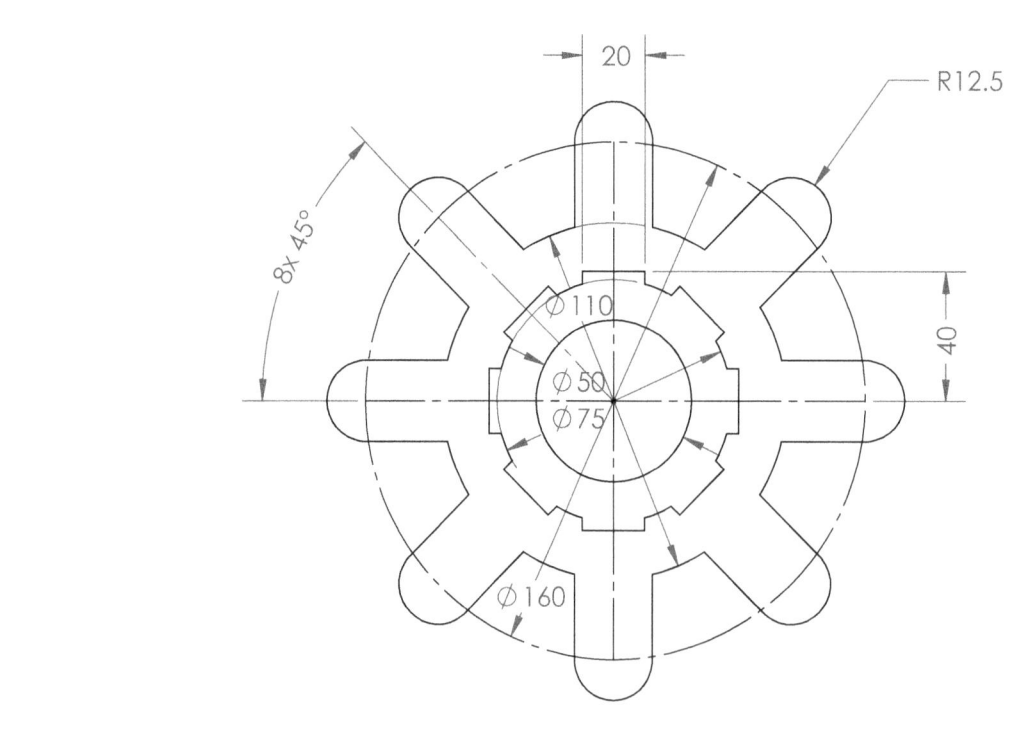

ALL DIMENSIONS ARE IN MM

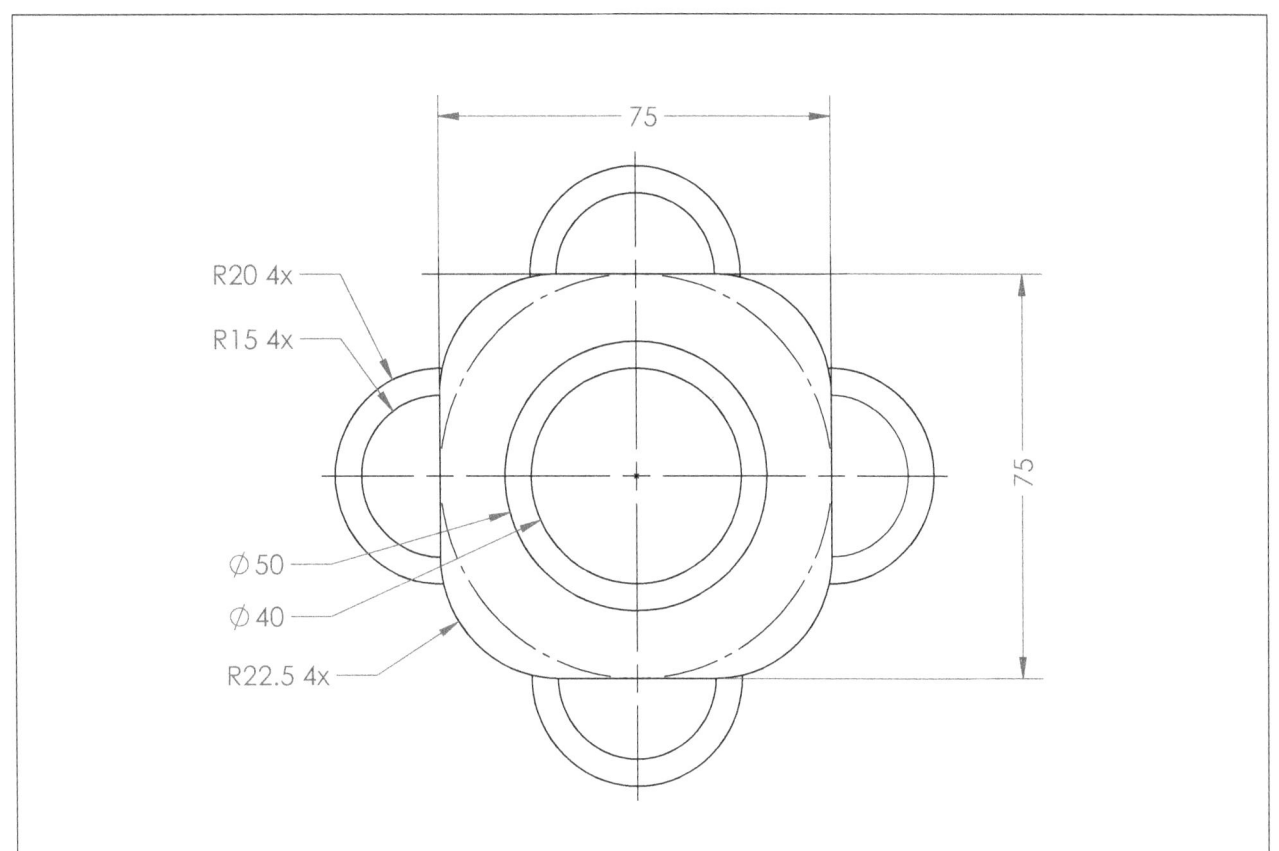
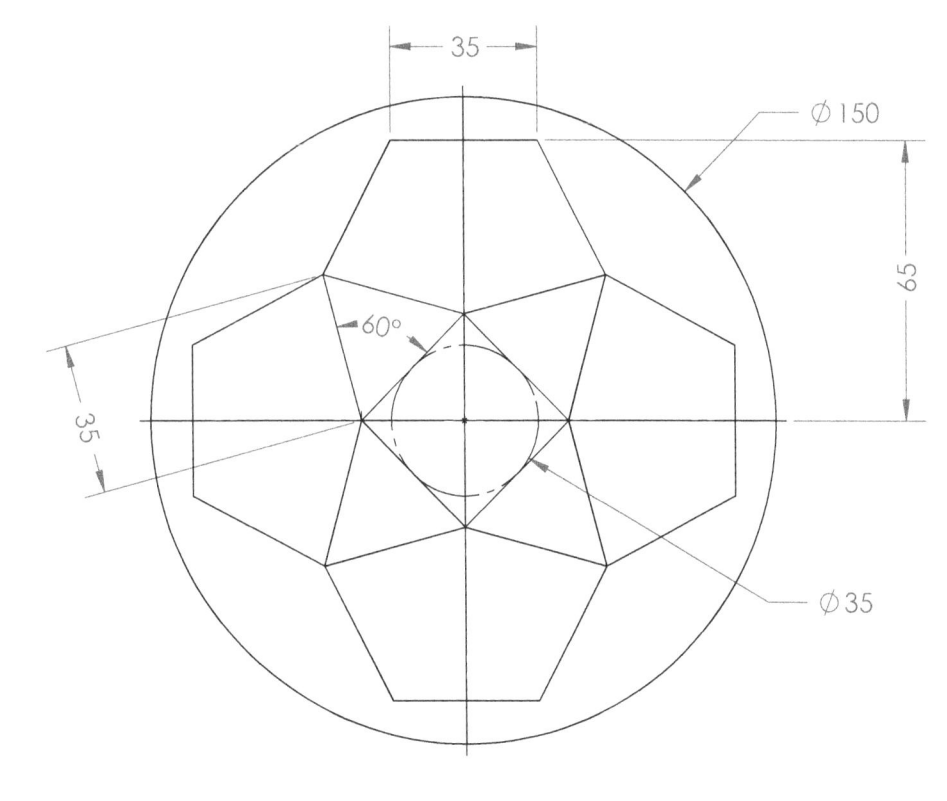

ALL DIMENSIONS ARE IN MM

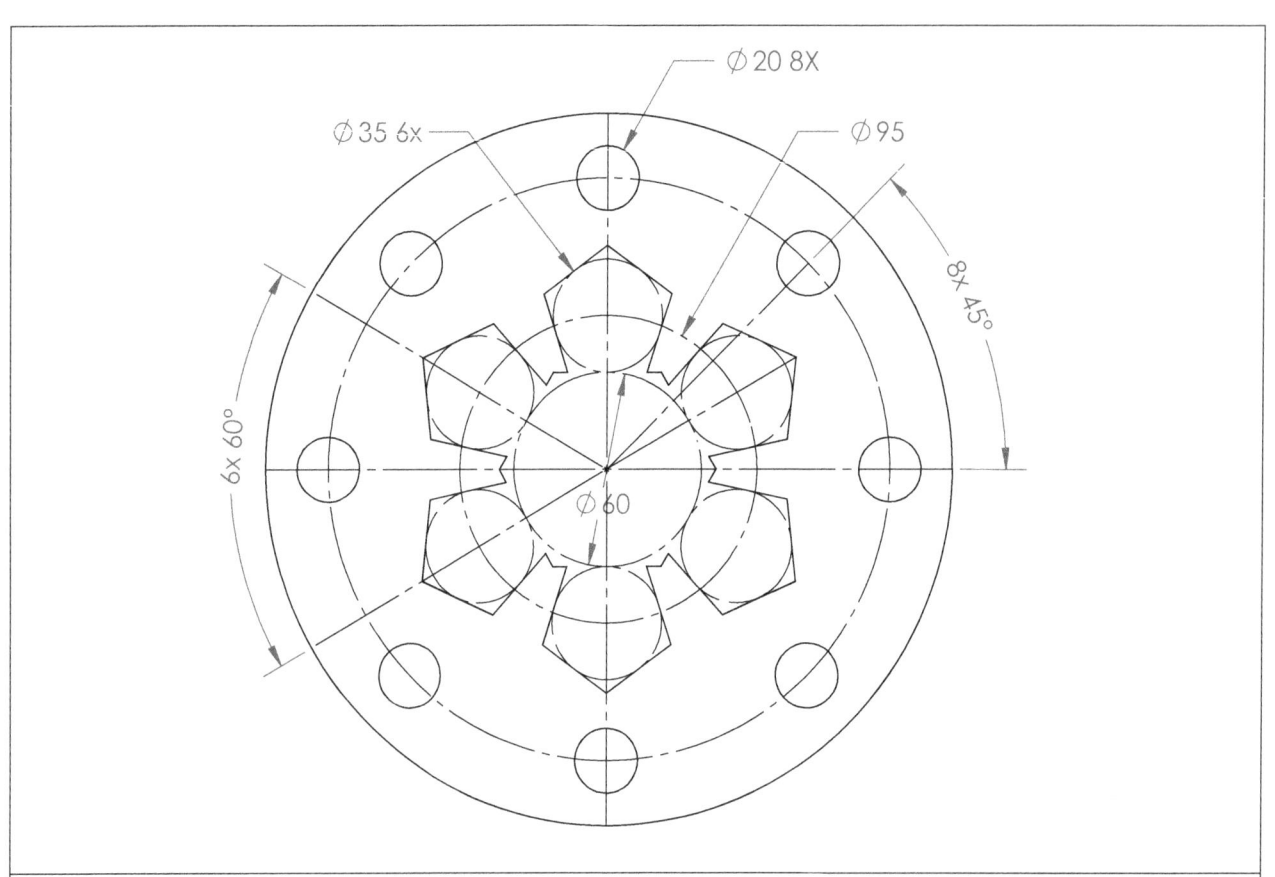
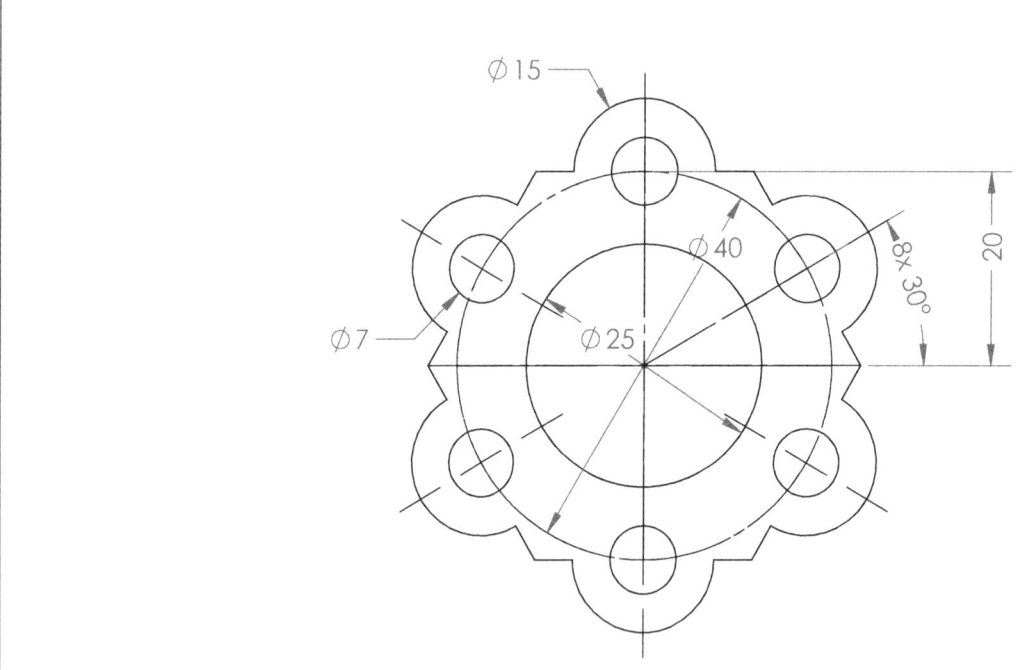

ALL DIMENSIONS ARE IN MM

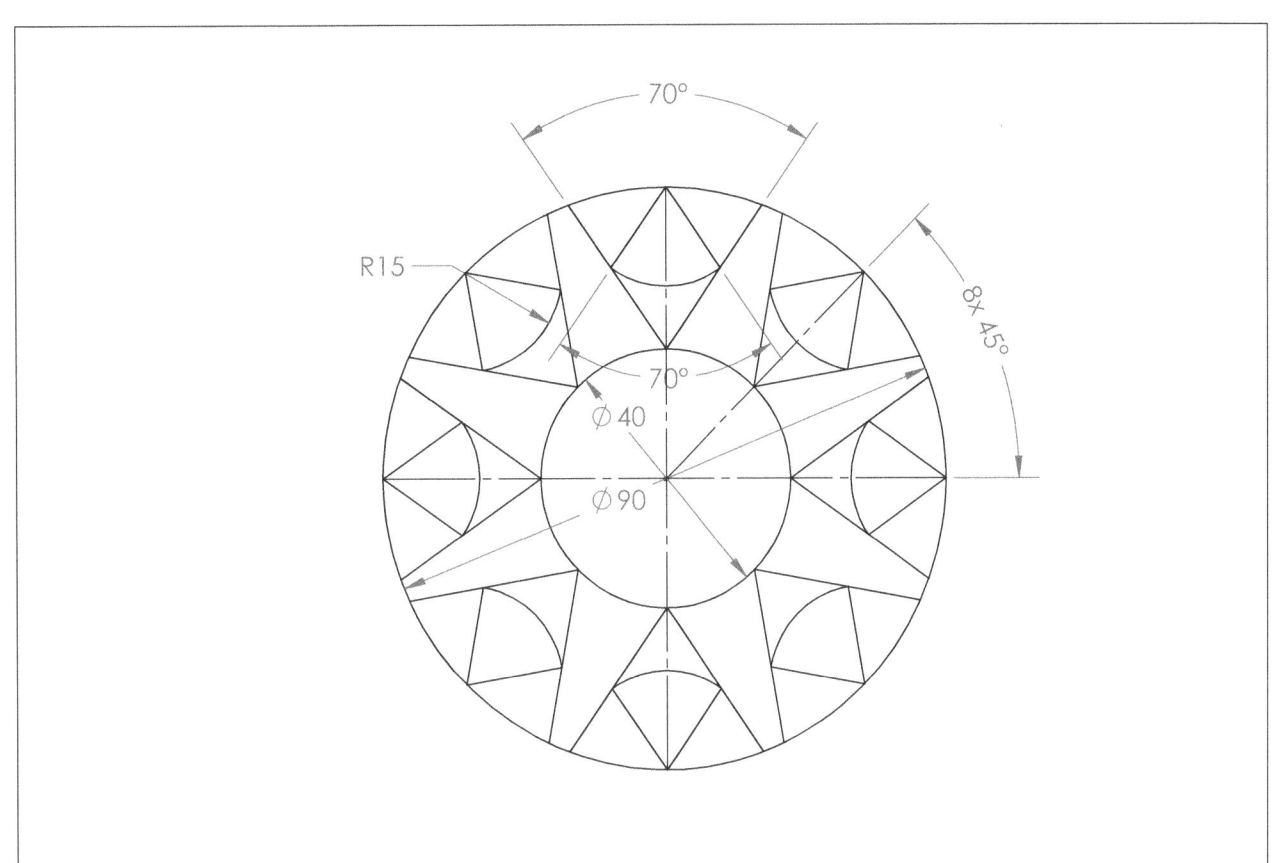
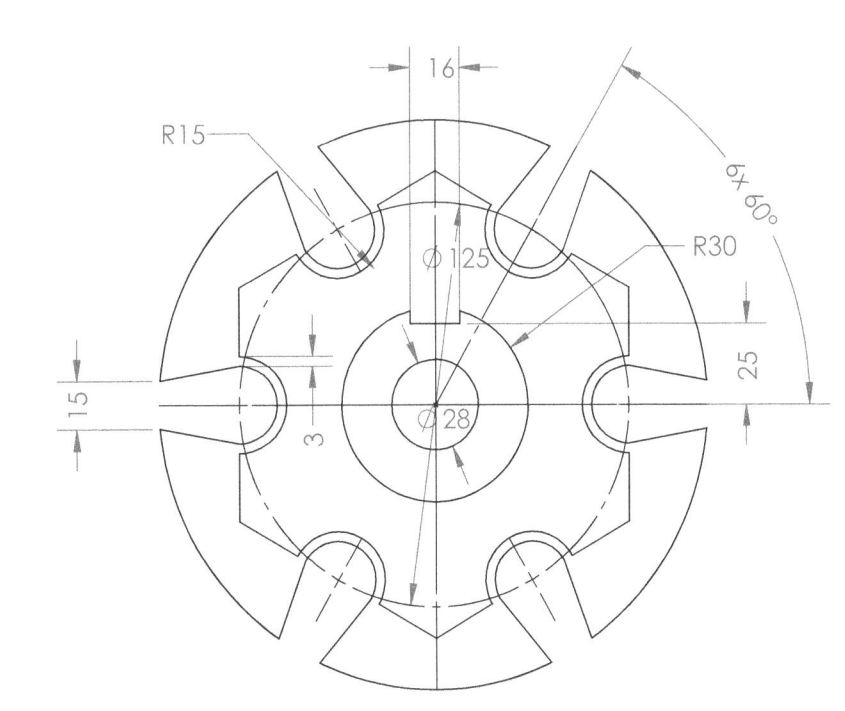

ALL DIMENSIONS ARE IN MM

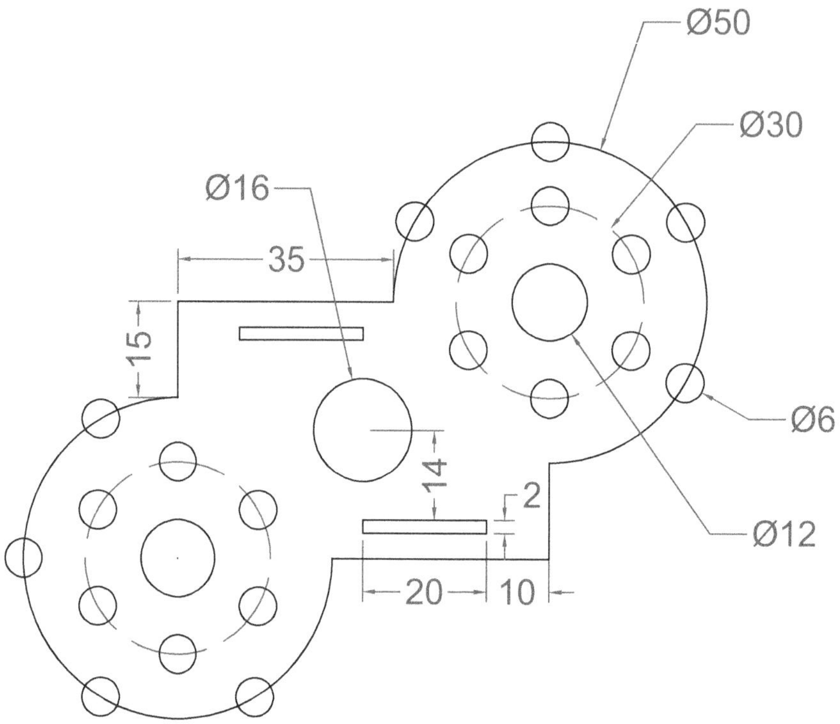

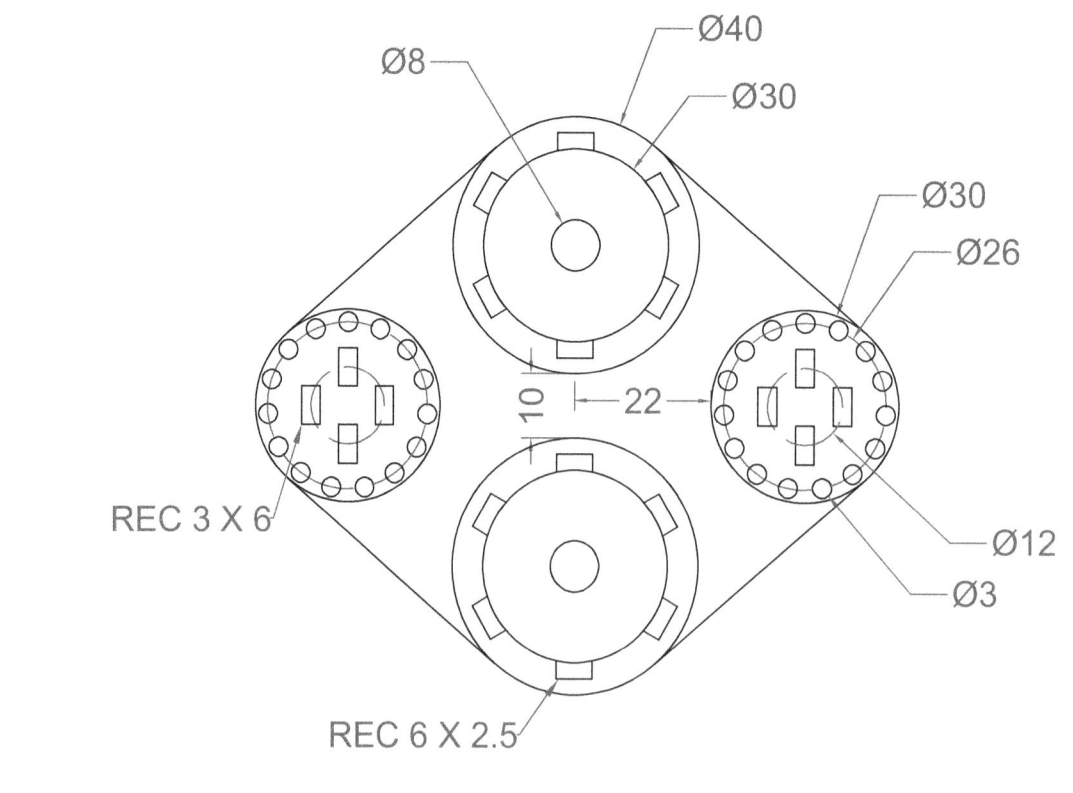

ALL DIMENSIONS ARE IN MM

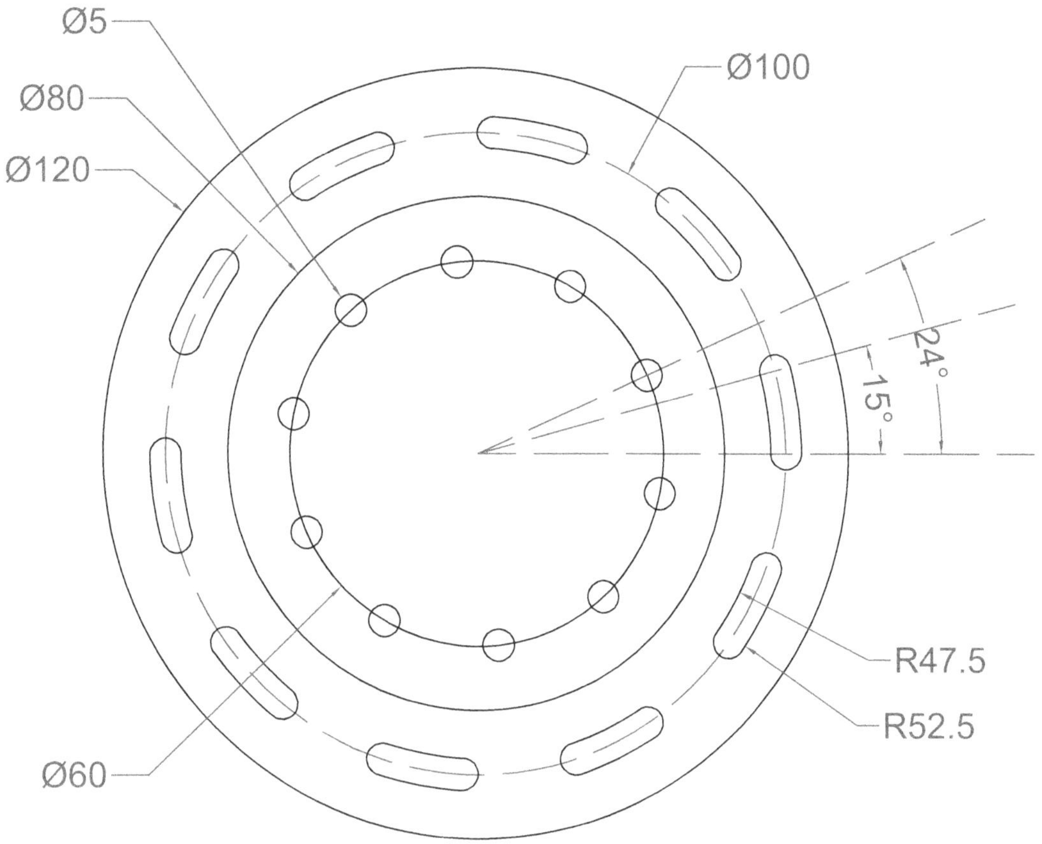

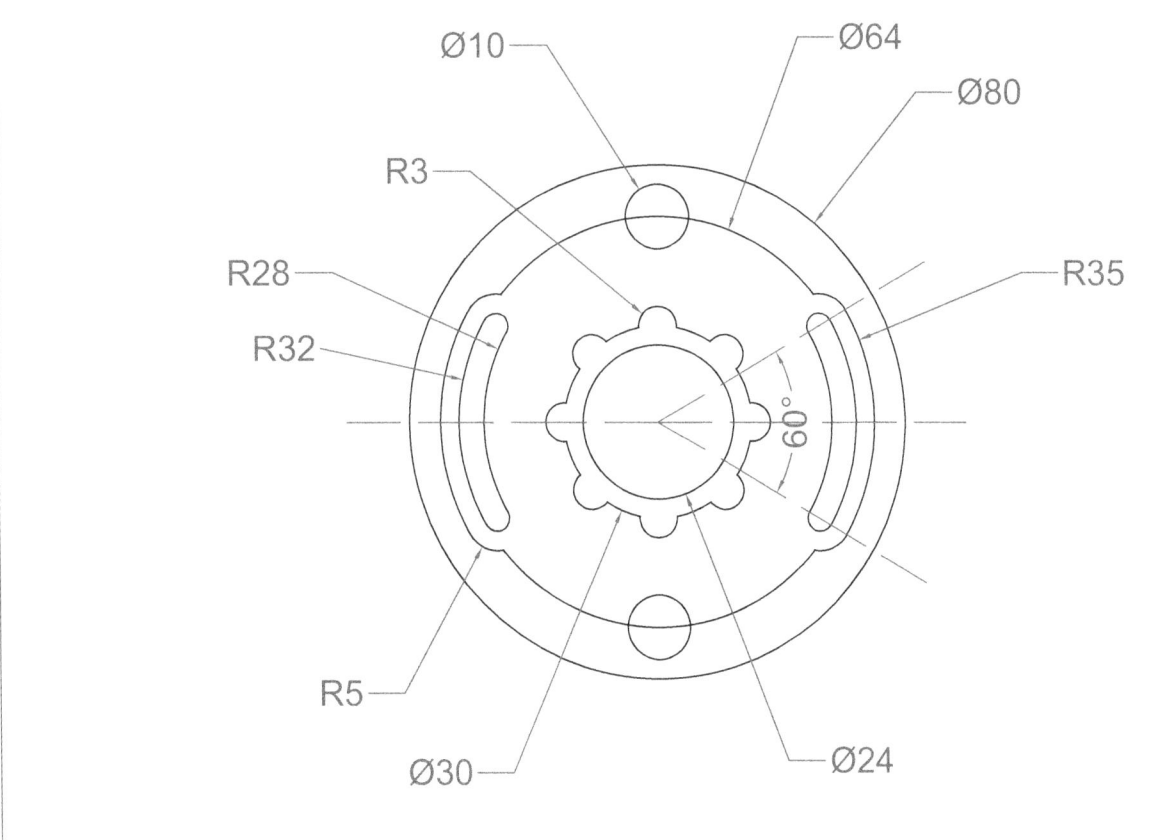

ALL DIMENSIONS ARE IN MM

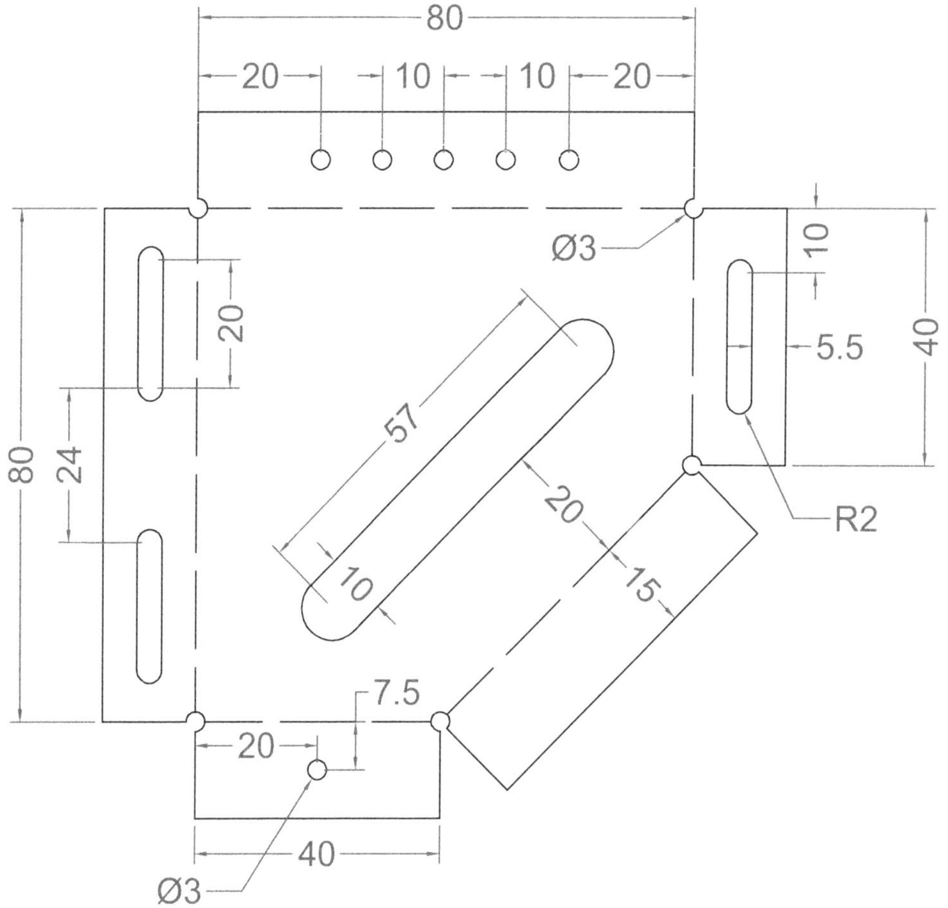

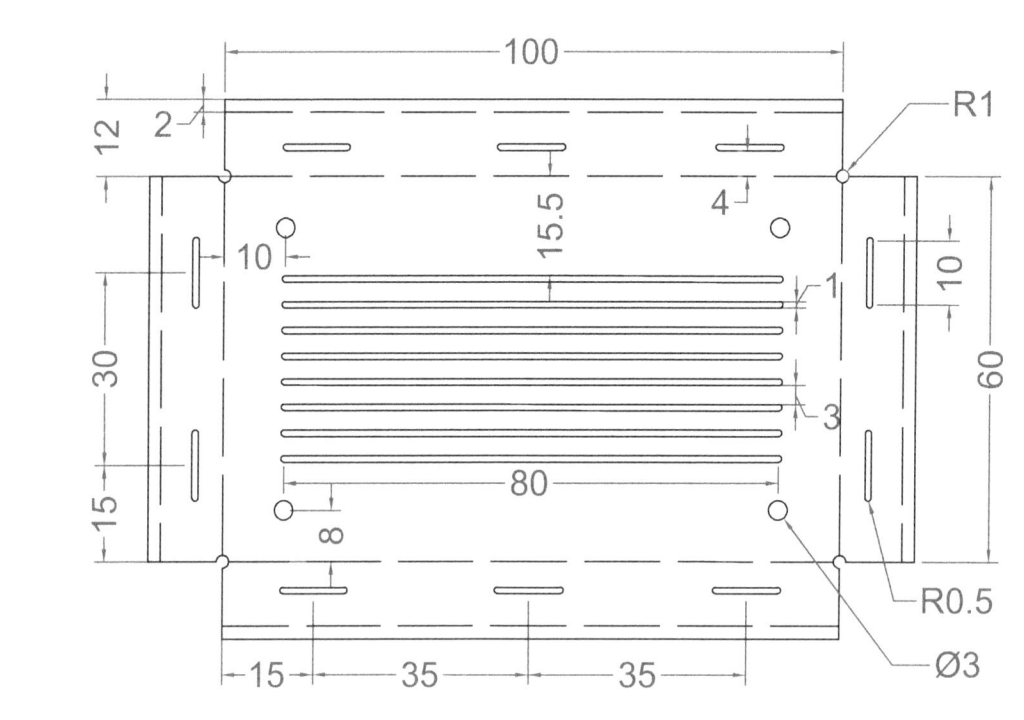

ALL DIMENSIONS ARE IN MM

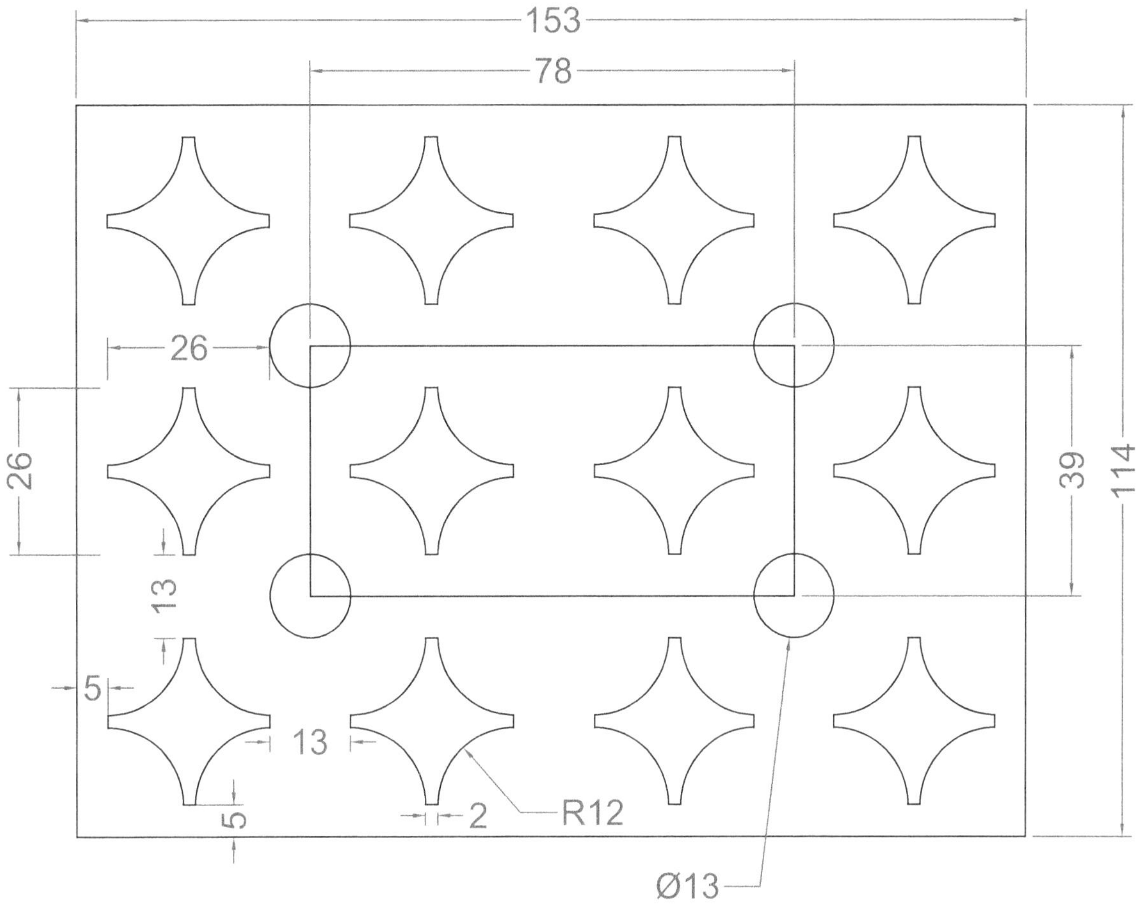

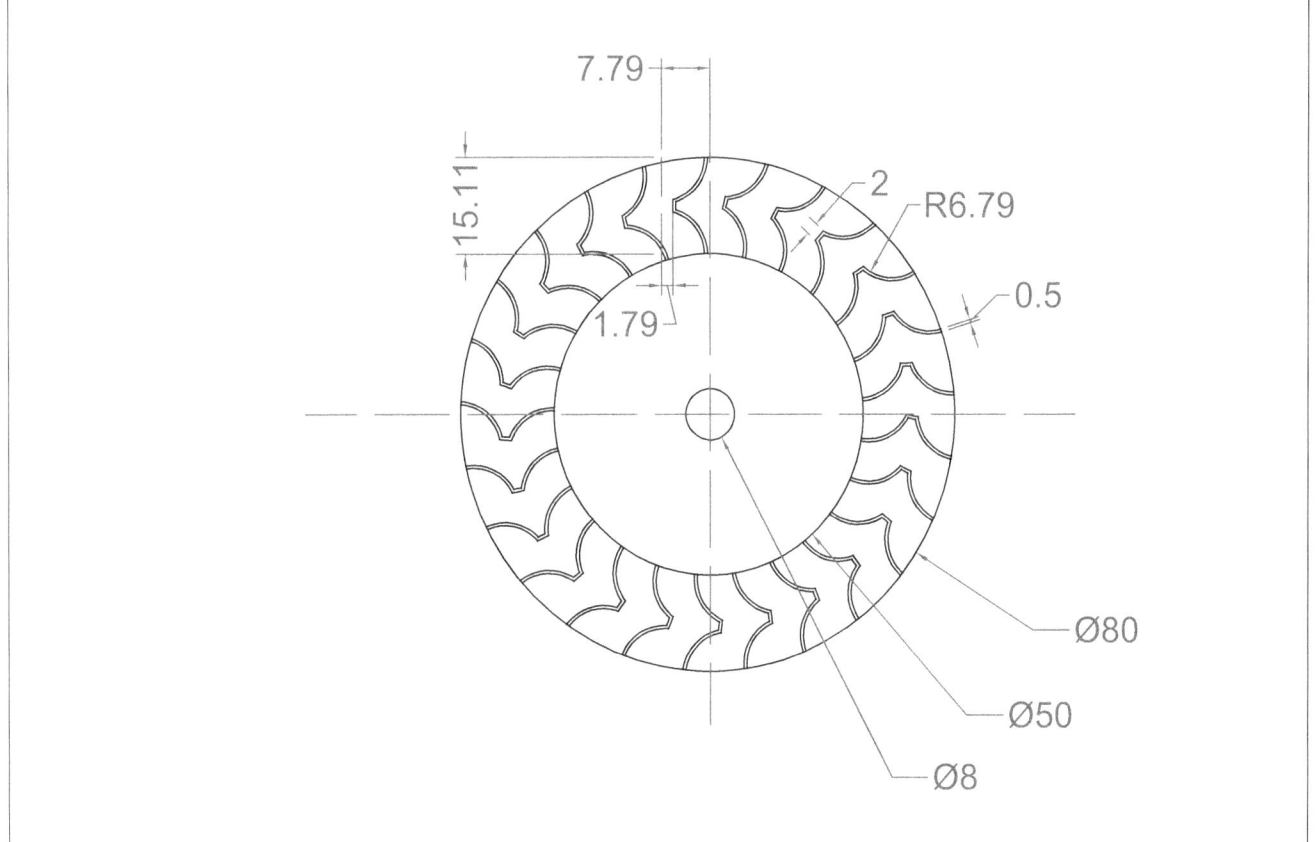

ALL DIMENSIONS ARE IN MM

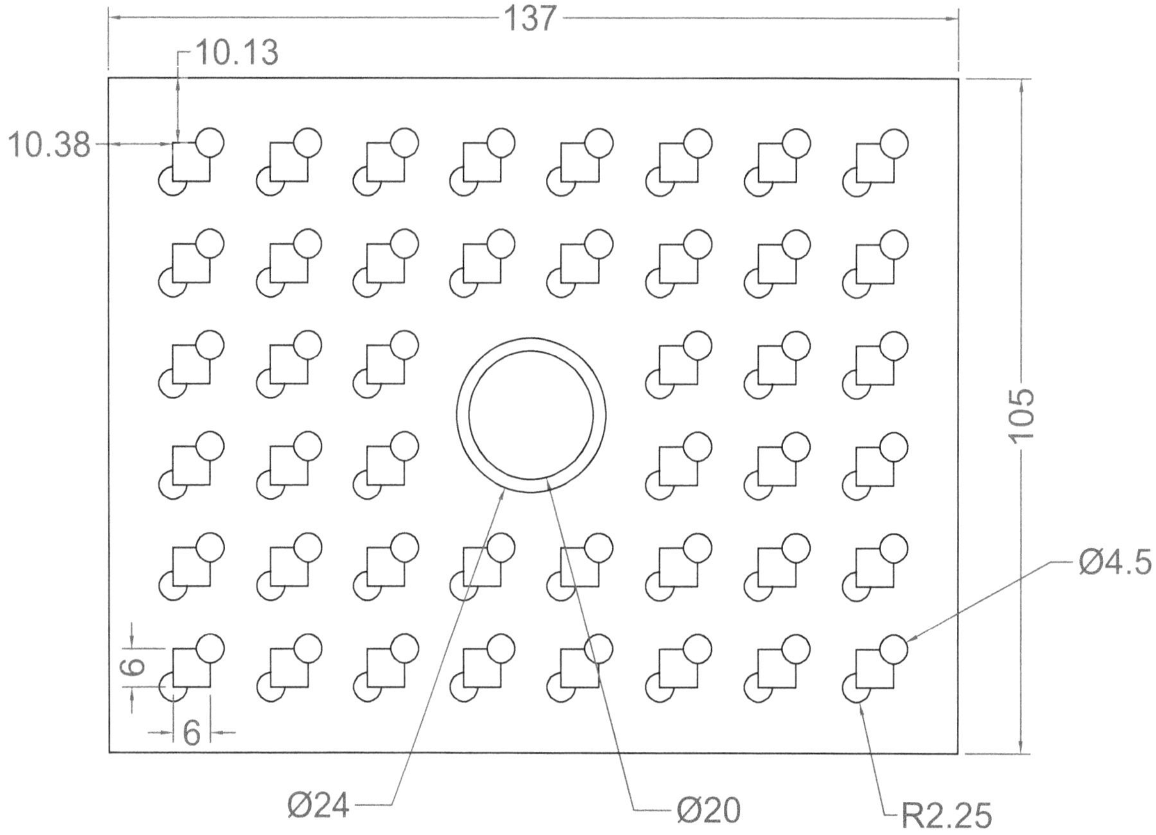

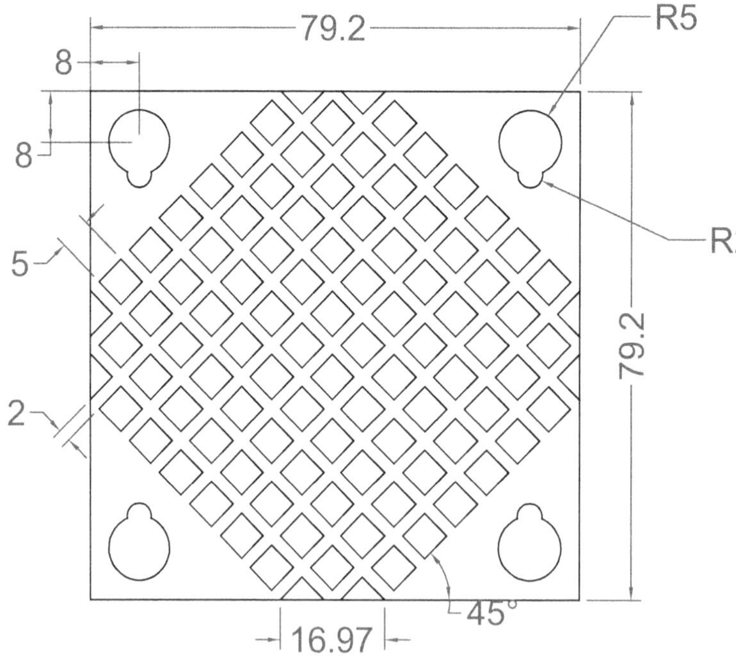

ALL DIMENSIONS ARE IN MM

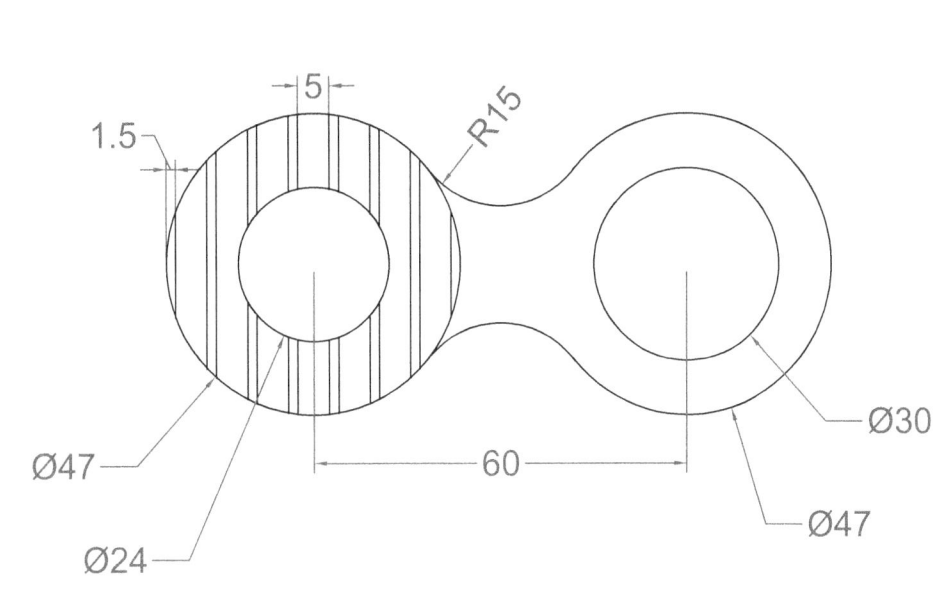
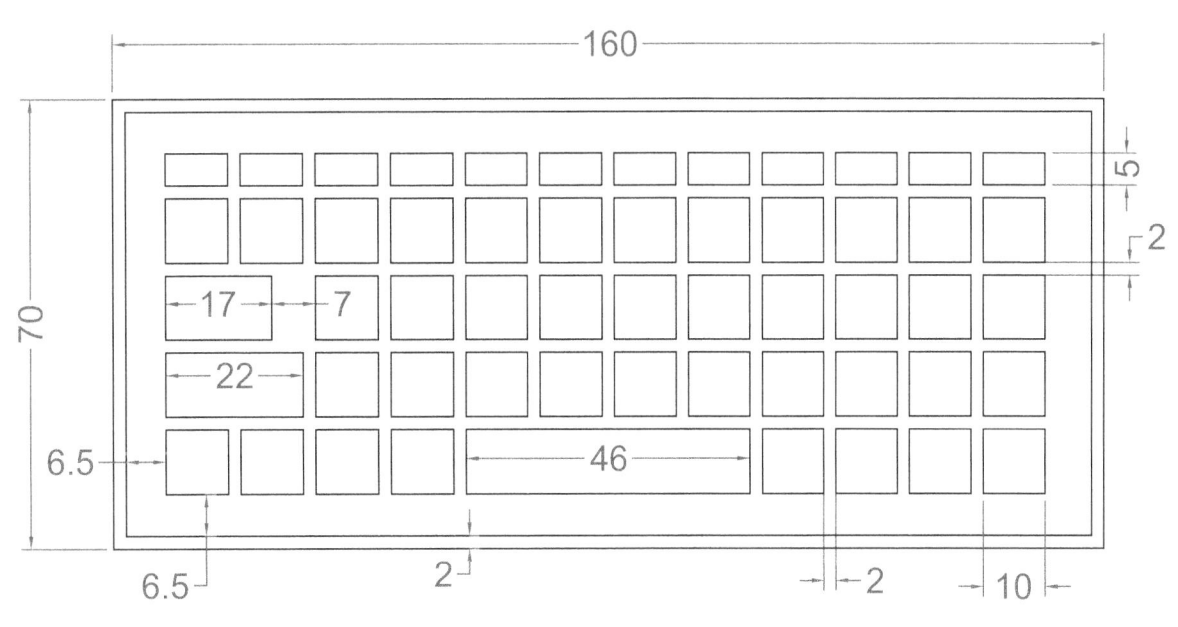

ALL DIMENSIONS ARE IN MM

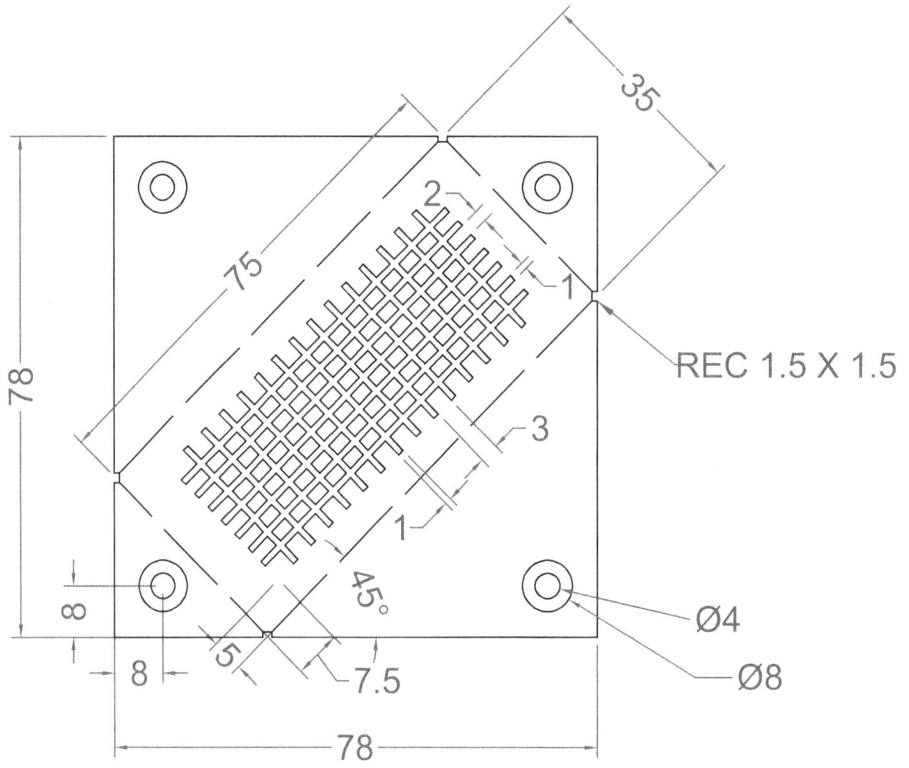

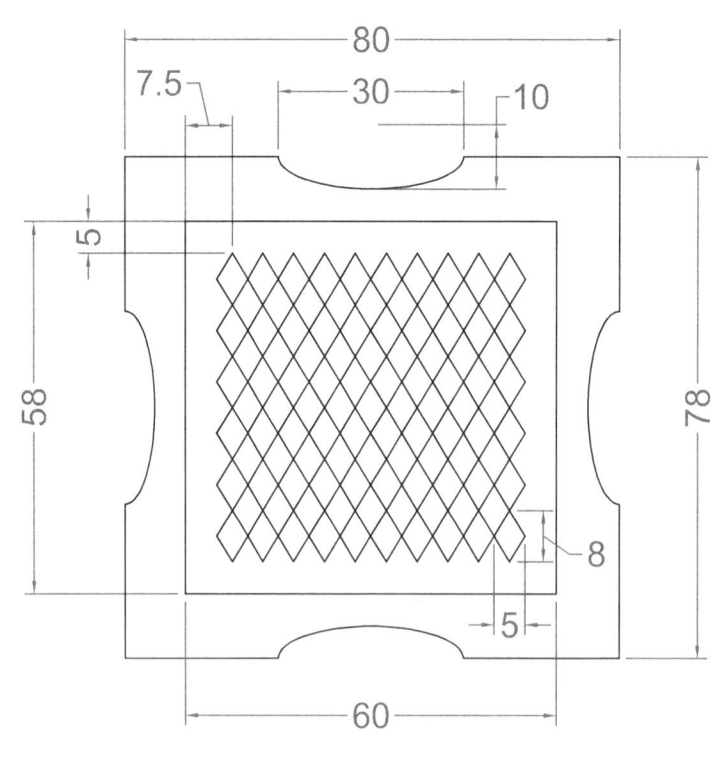

ALL DIMENSIONS ARE IN MM

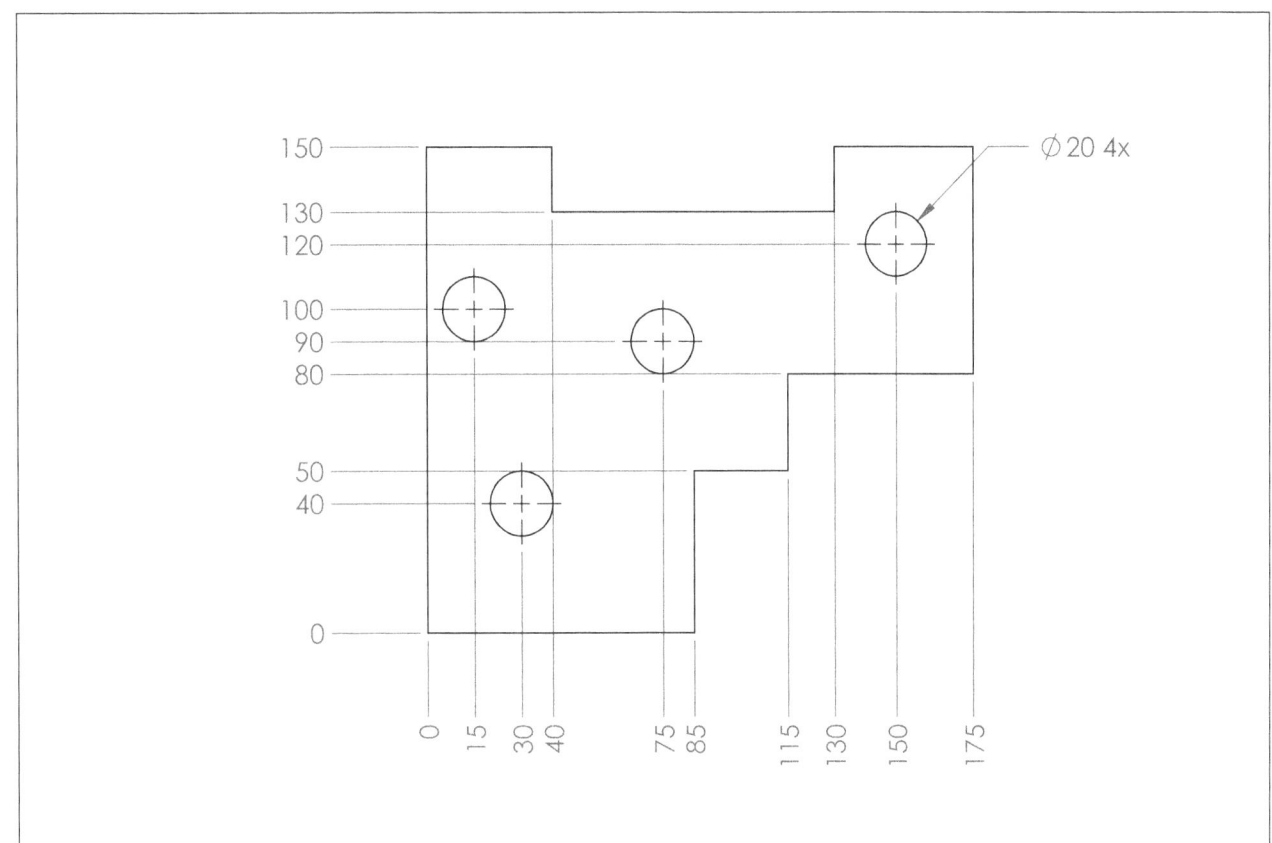
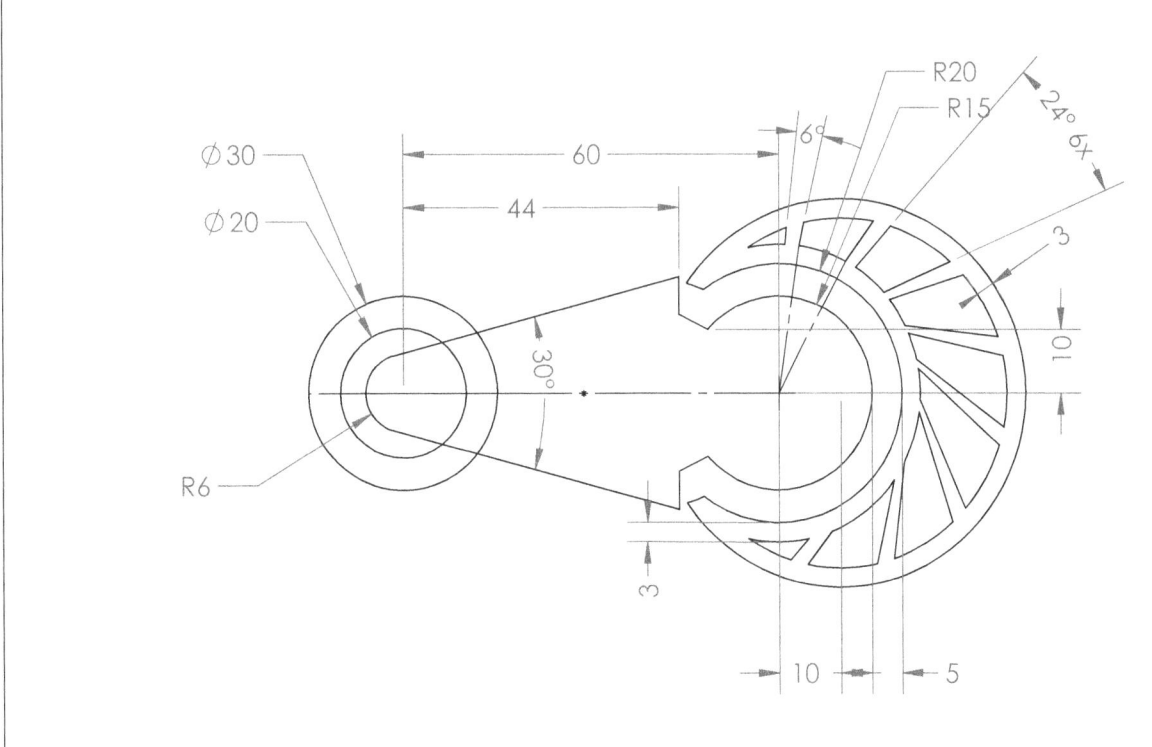

ALL DIMENSIONS ARE IN MM

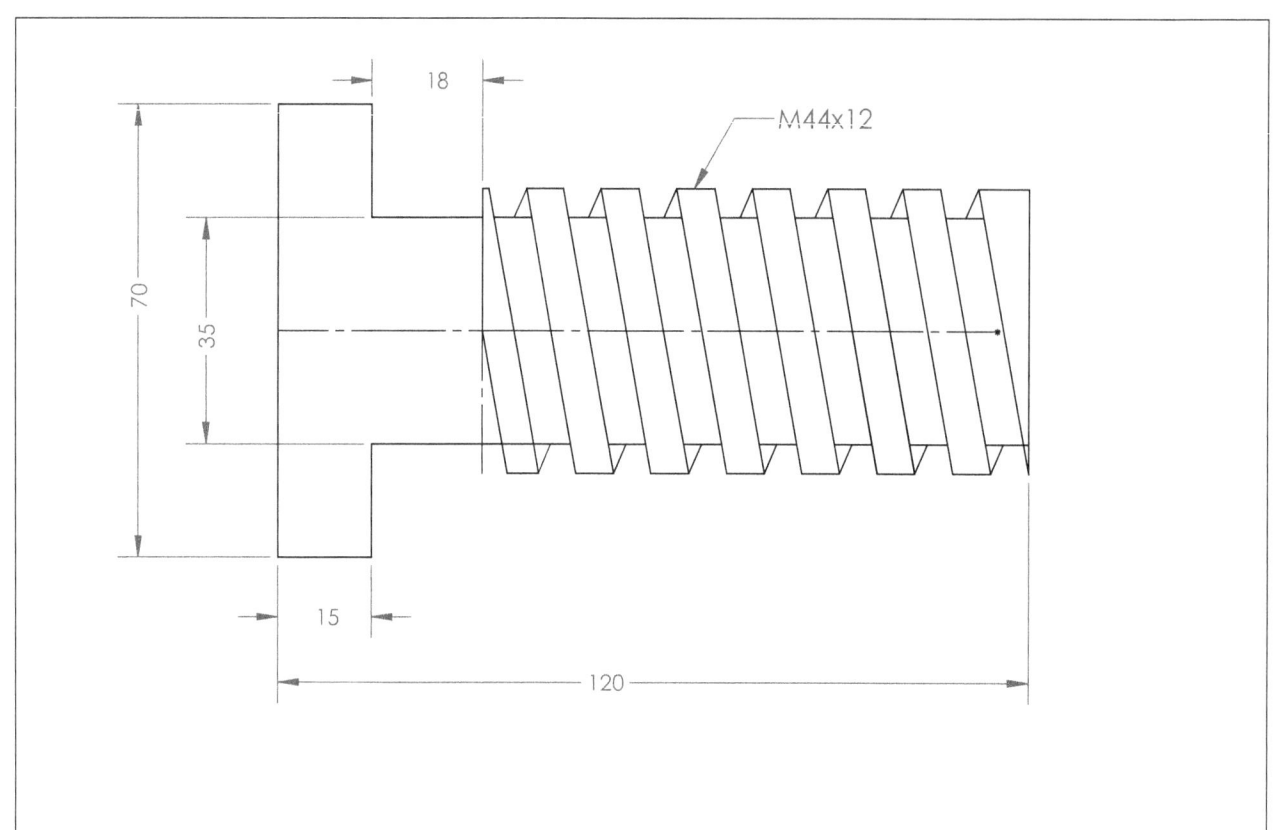

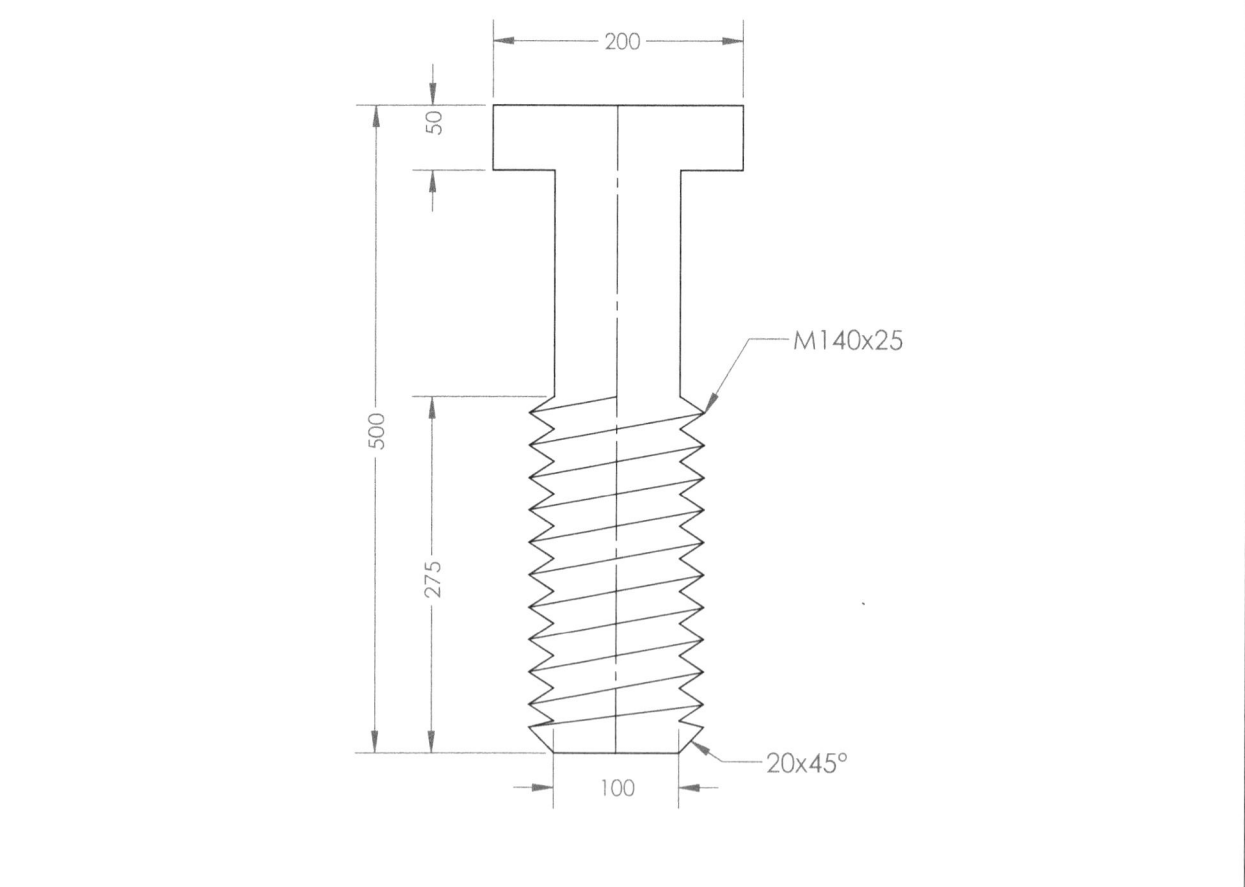

ALL DIMENSIONS ARE IN MM

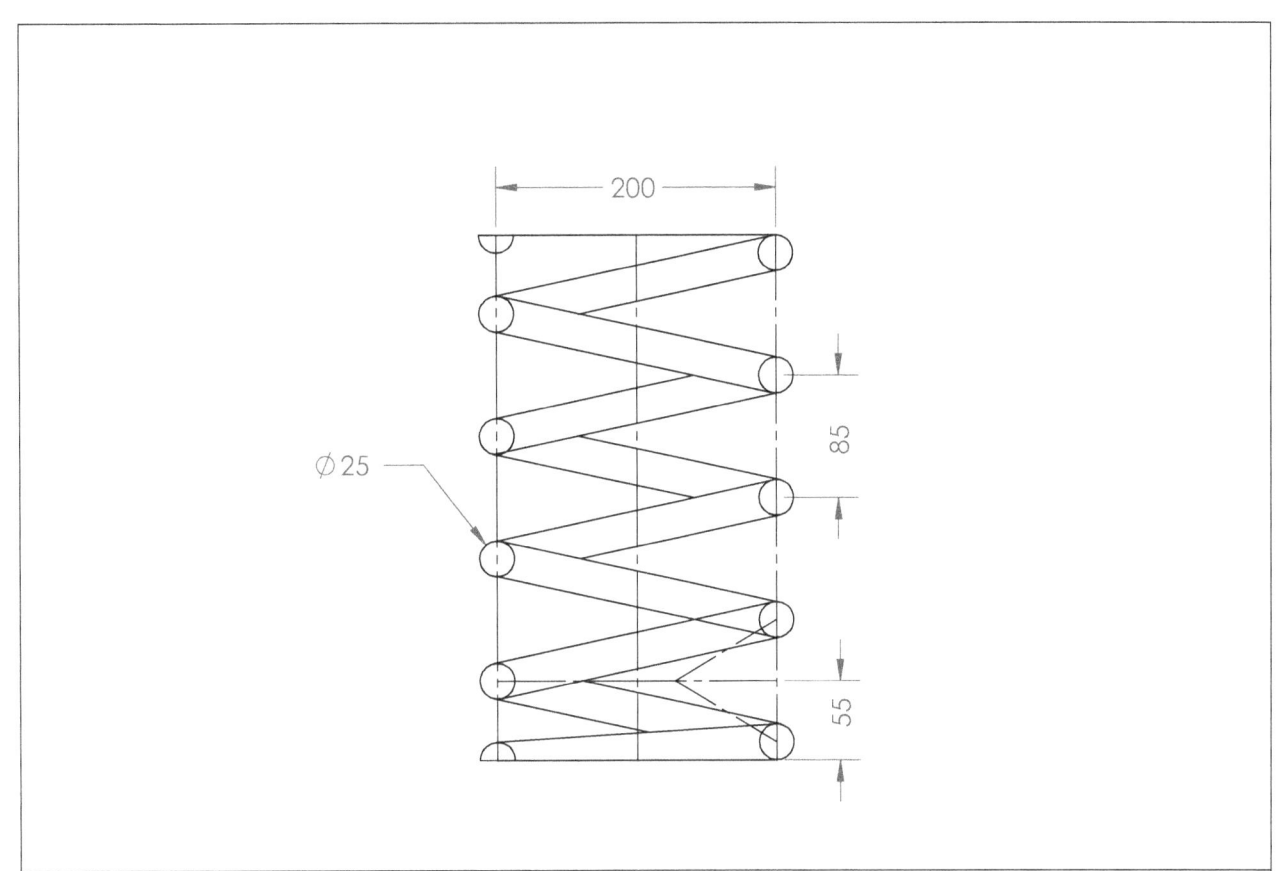
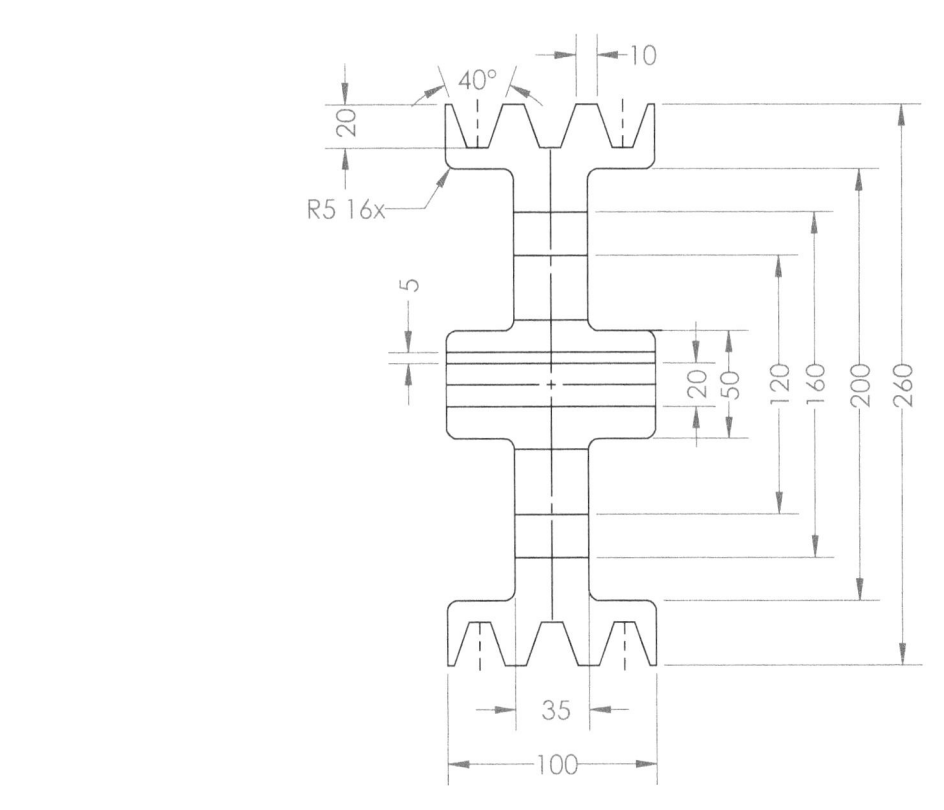

ALL DIMENSIONS ARE IN MM

3D CAD EXERCISES

Page intentionally left blank

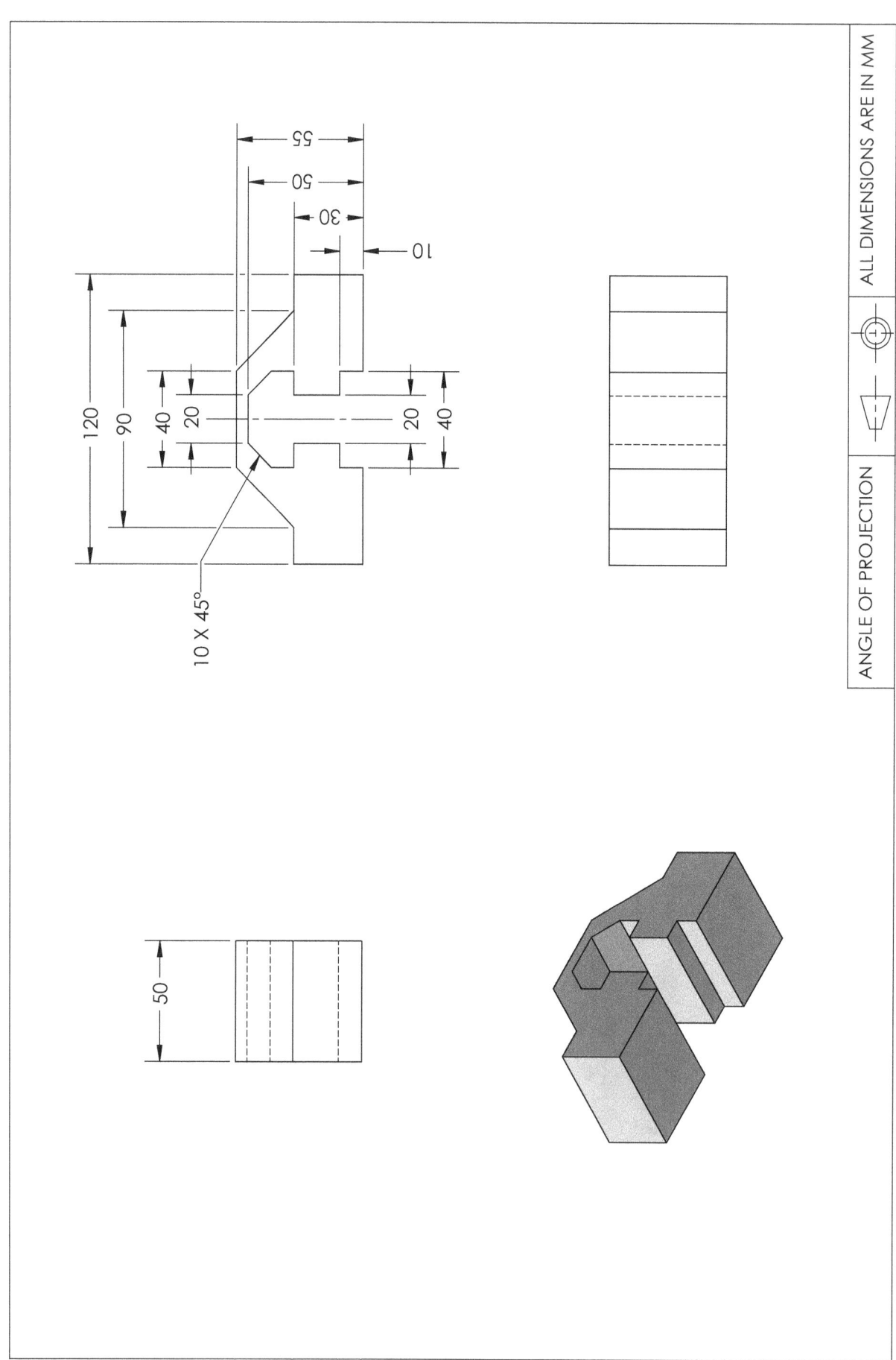

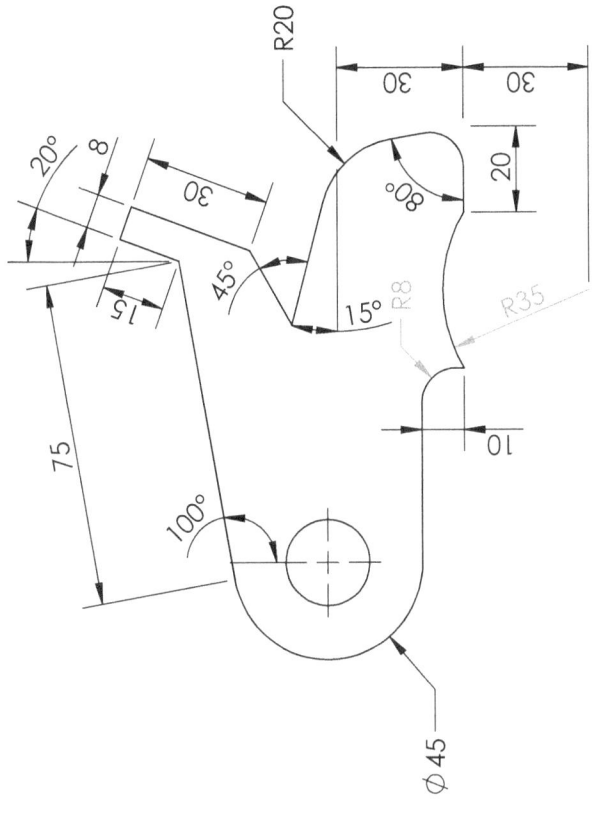
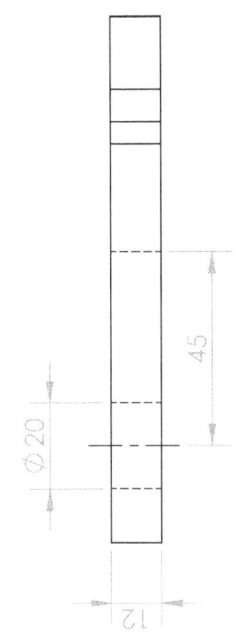
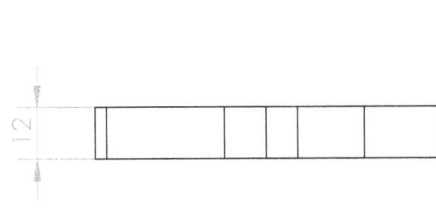
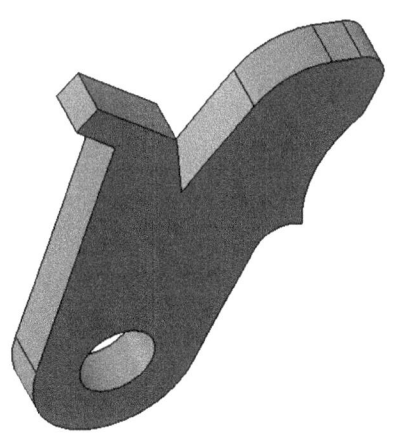

ALL DIMENSIONS ARE IN MM

ANGLE OF PROJECTION

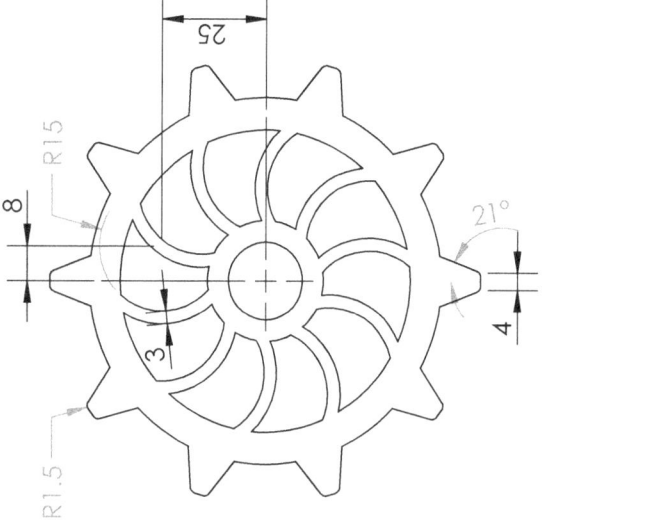
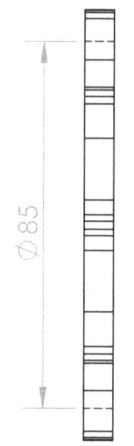
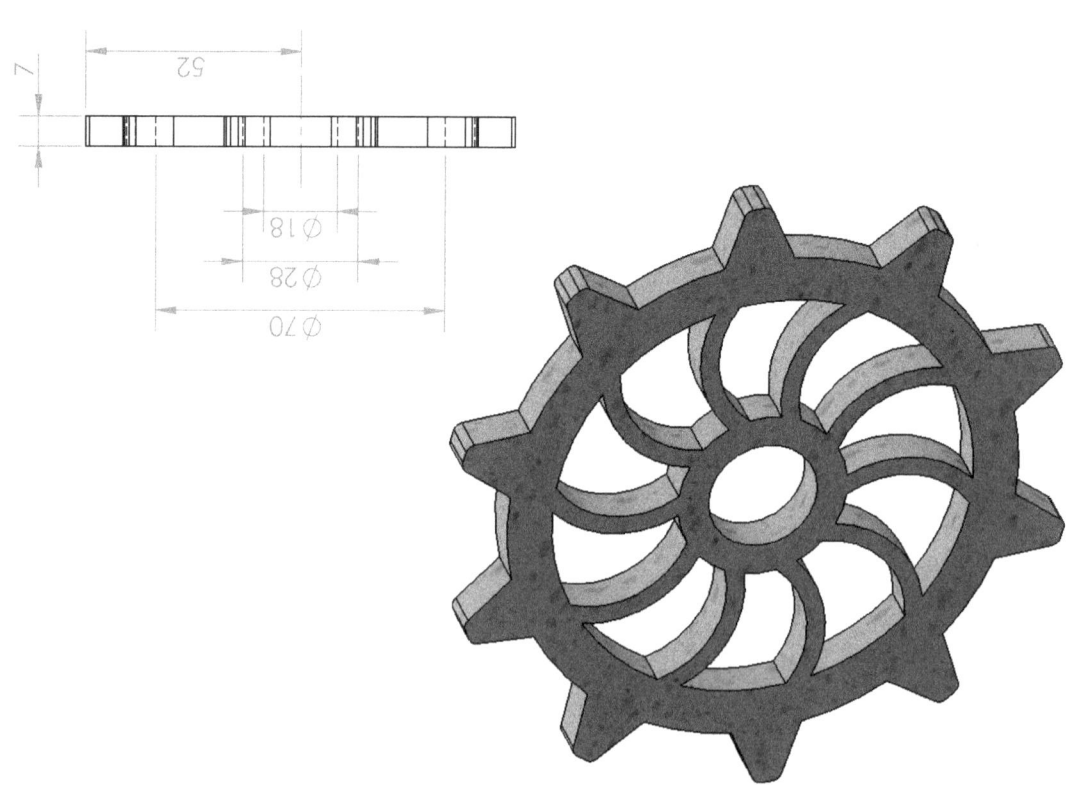

ALL DIMENSIONS ARE IN MM

ANGLE OF PROJECTION

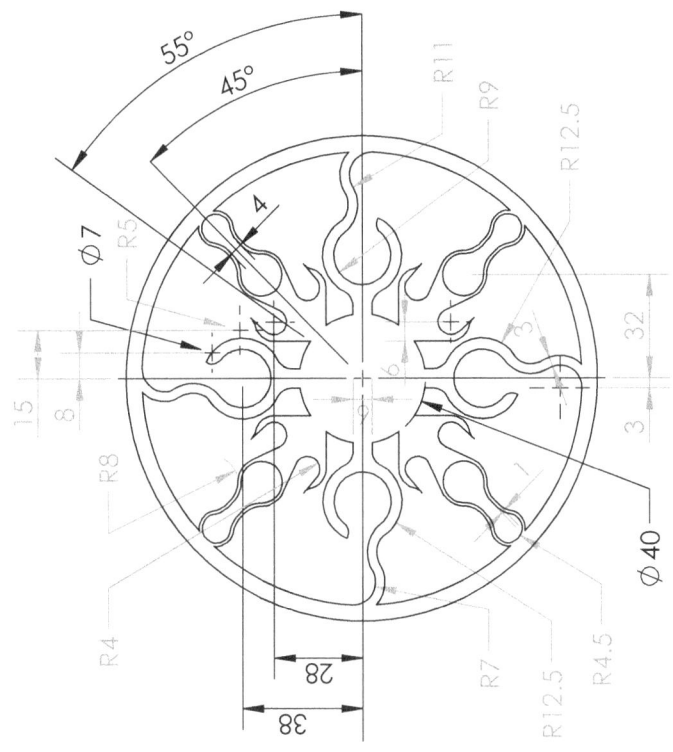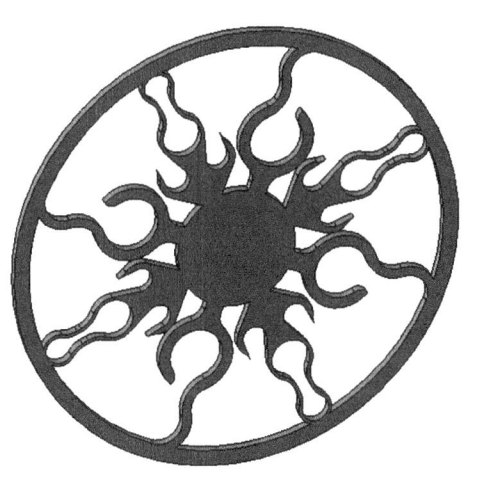

ALL DIMENSIONS ARE IN MM

ANGLE OF PROJECTION

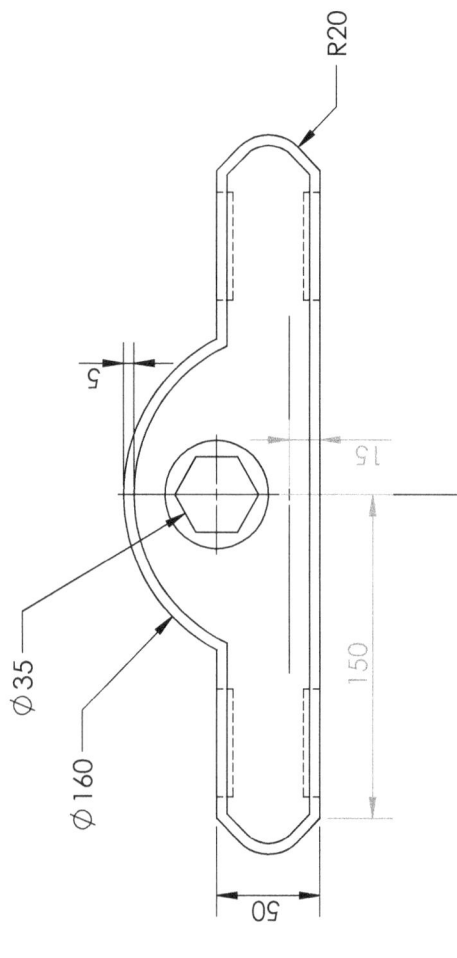
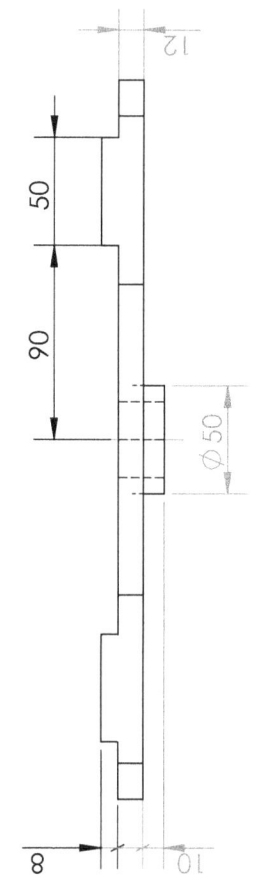
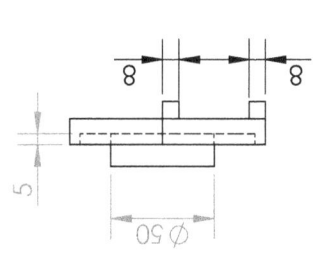
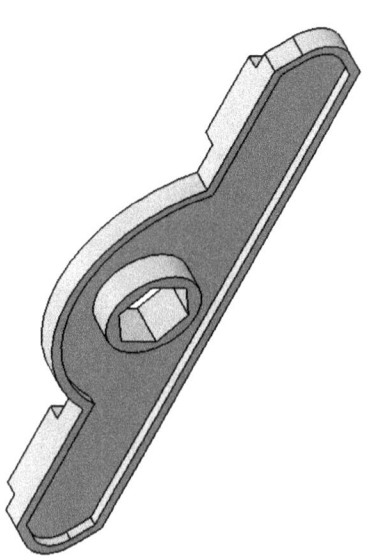

ALL DIMENSIONS ARE IN MM

ANGLE OF PROJECTION

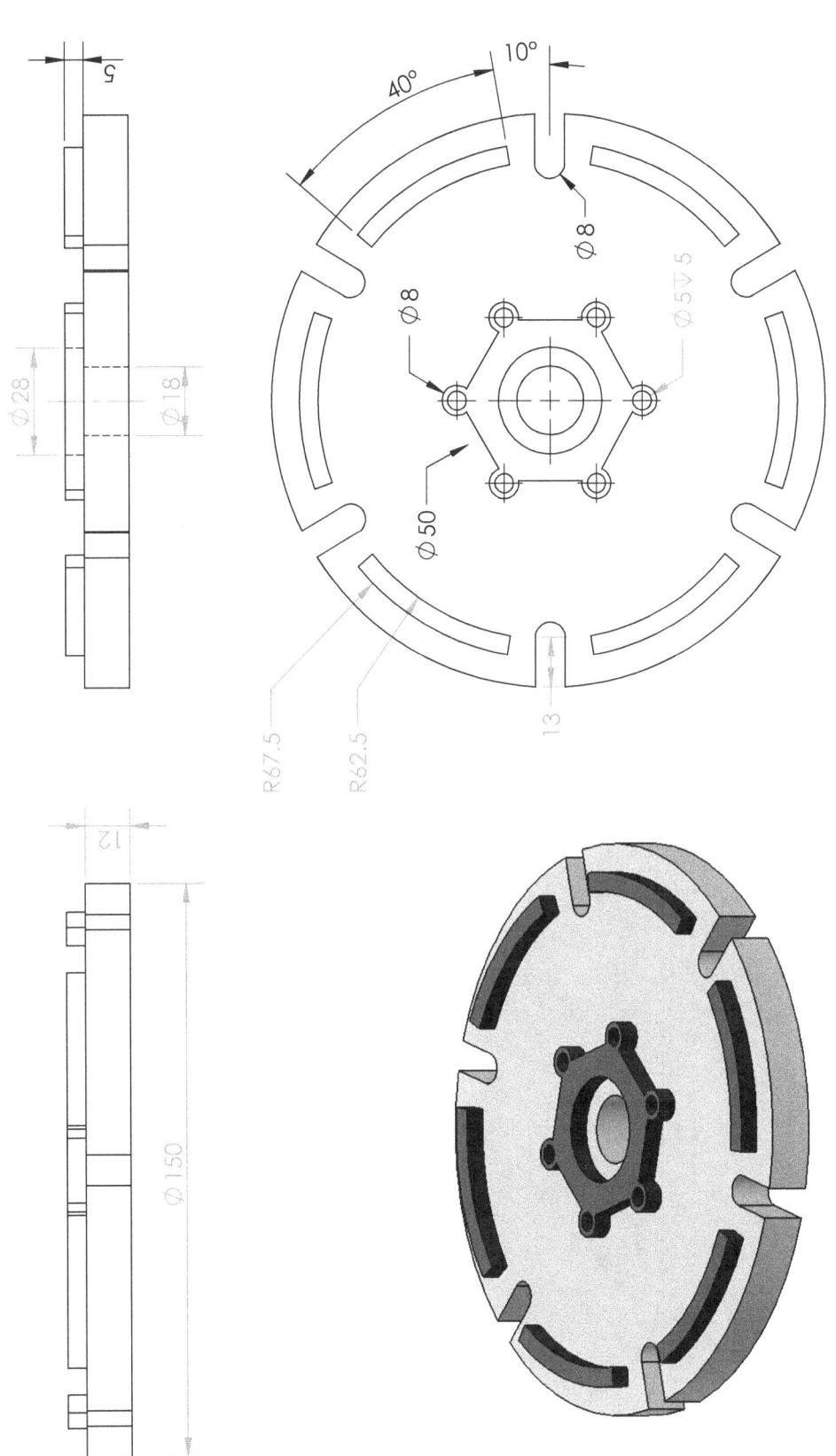

ALL DIMENSIONS ARE IN MM

ANGLE OF PROJECTION

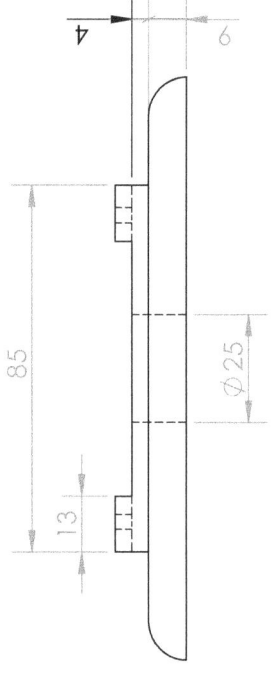
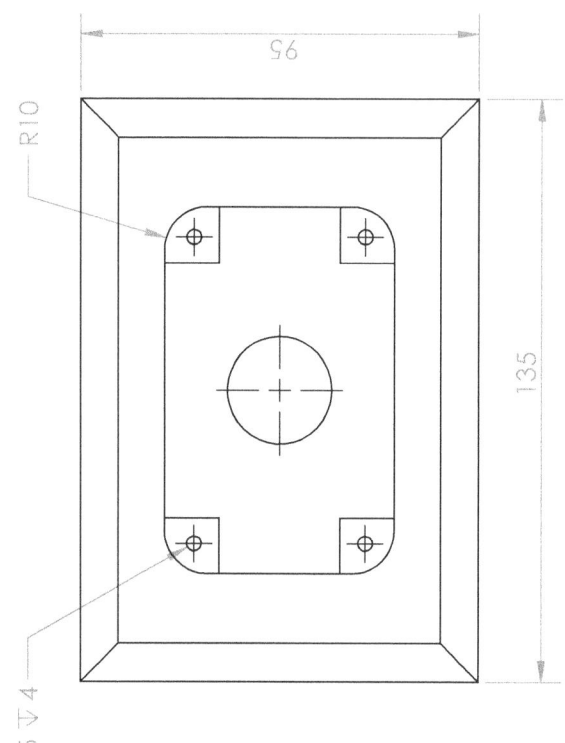
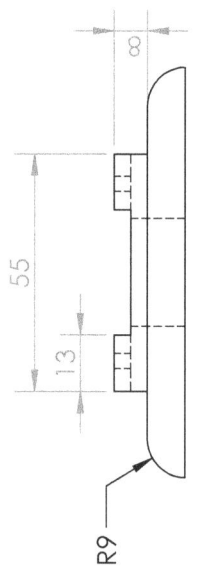
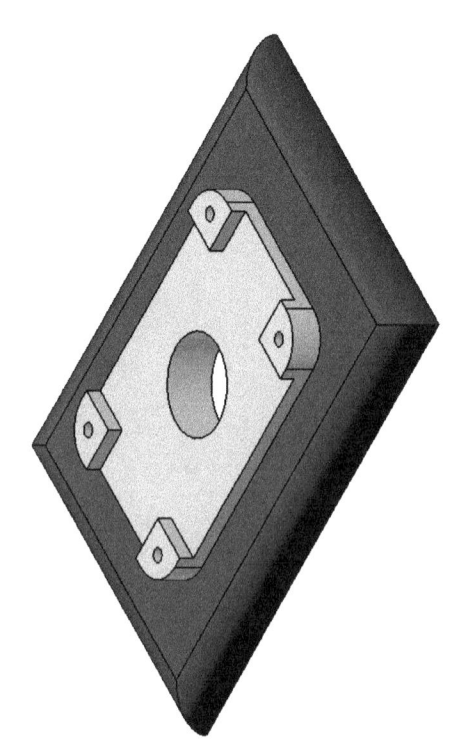

ALL DIMENSIONS ARE IN MM

ANGLE OF PROJECTION

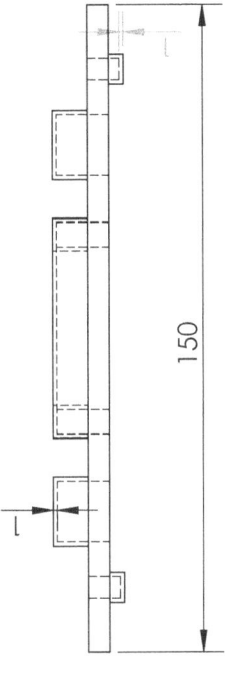
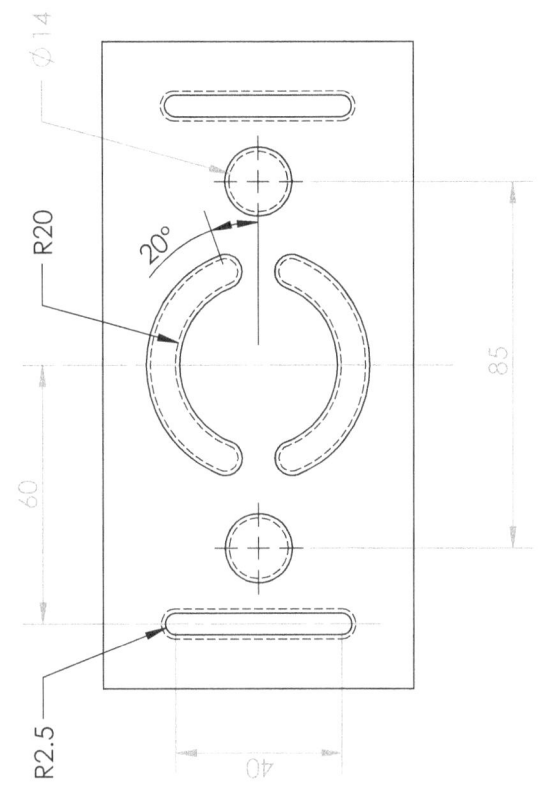
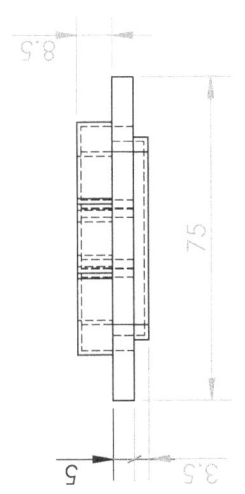
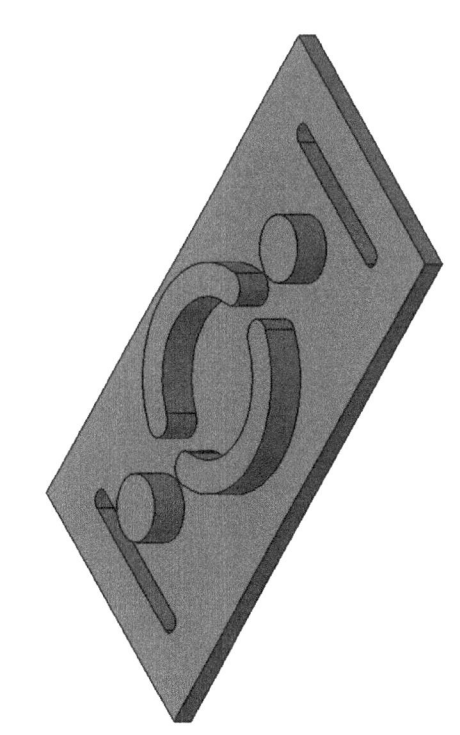

ALL DIMENSIONS ARE IN MM

ANGLE OF PROJECTION **EX :**

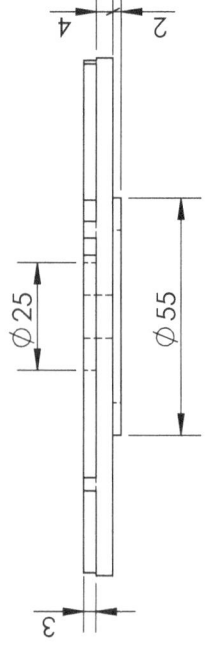
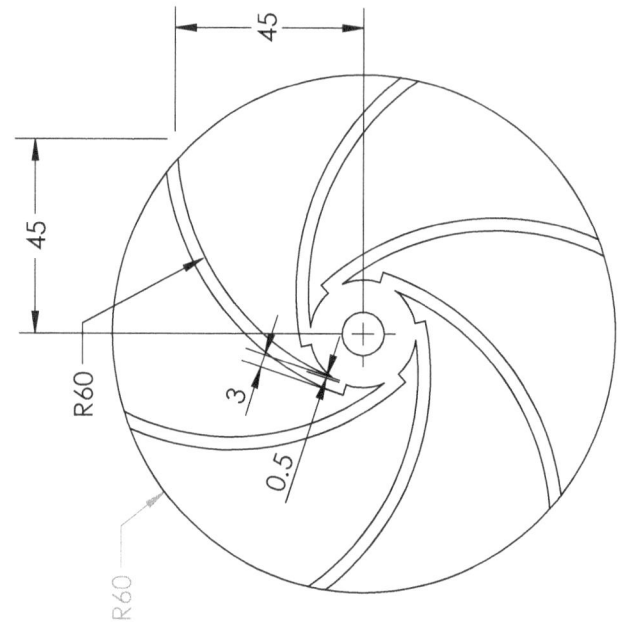
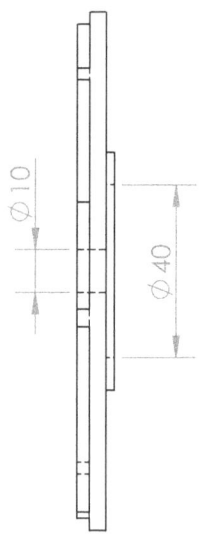
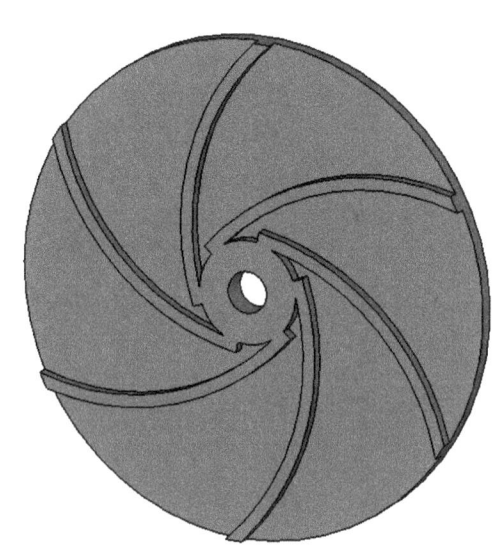

ALL DIMENSIONS ARE IN MM

ANGLE OF PROJECTION

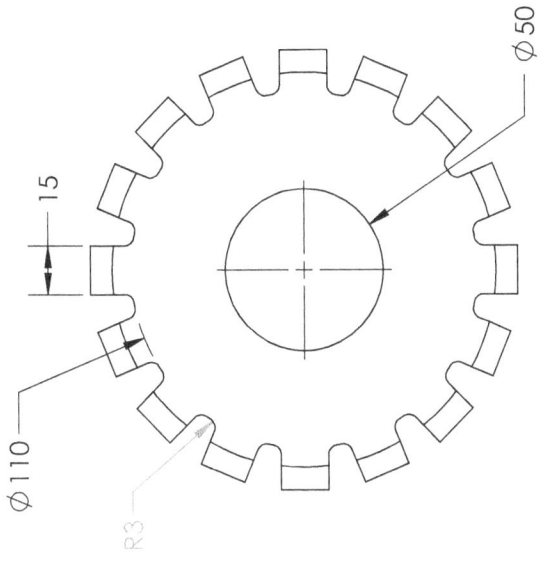
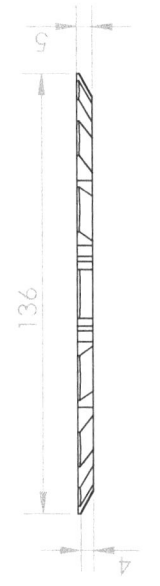
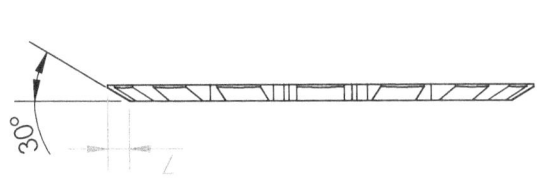

ALL DIMENSIONS ARE IN MM

ANGLE OF PROJECTION

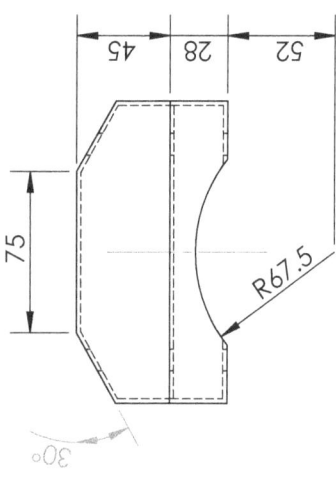
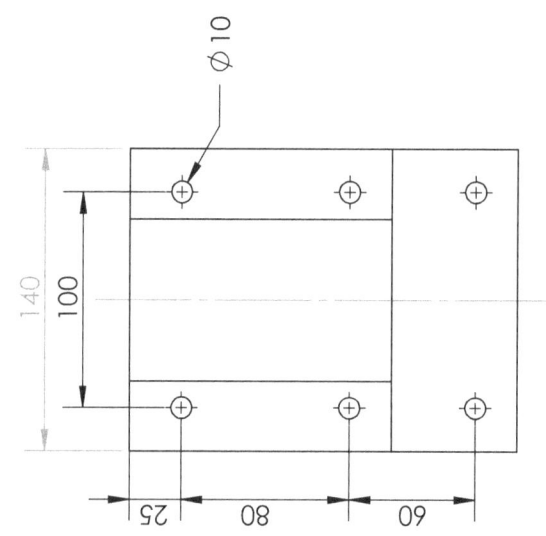
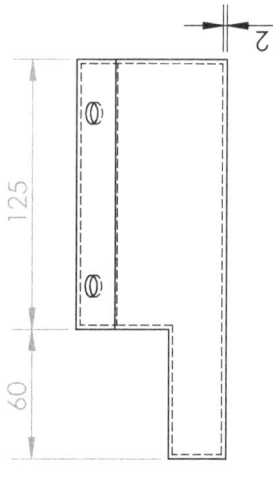
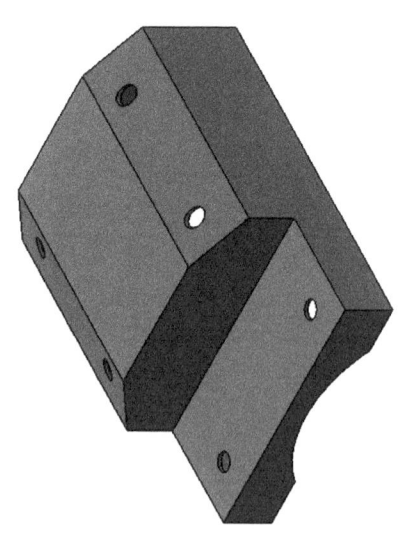

ALL DIMENSIONS ARE IN MM

ANGLE OF PROJECTION

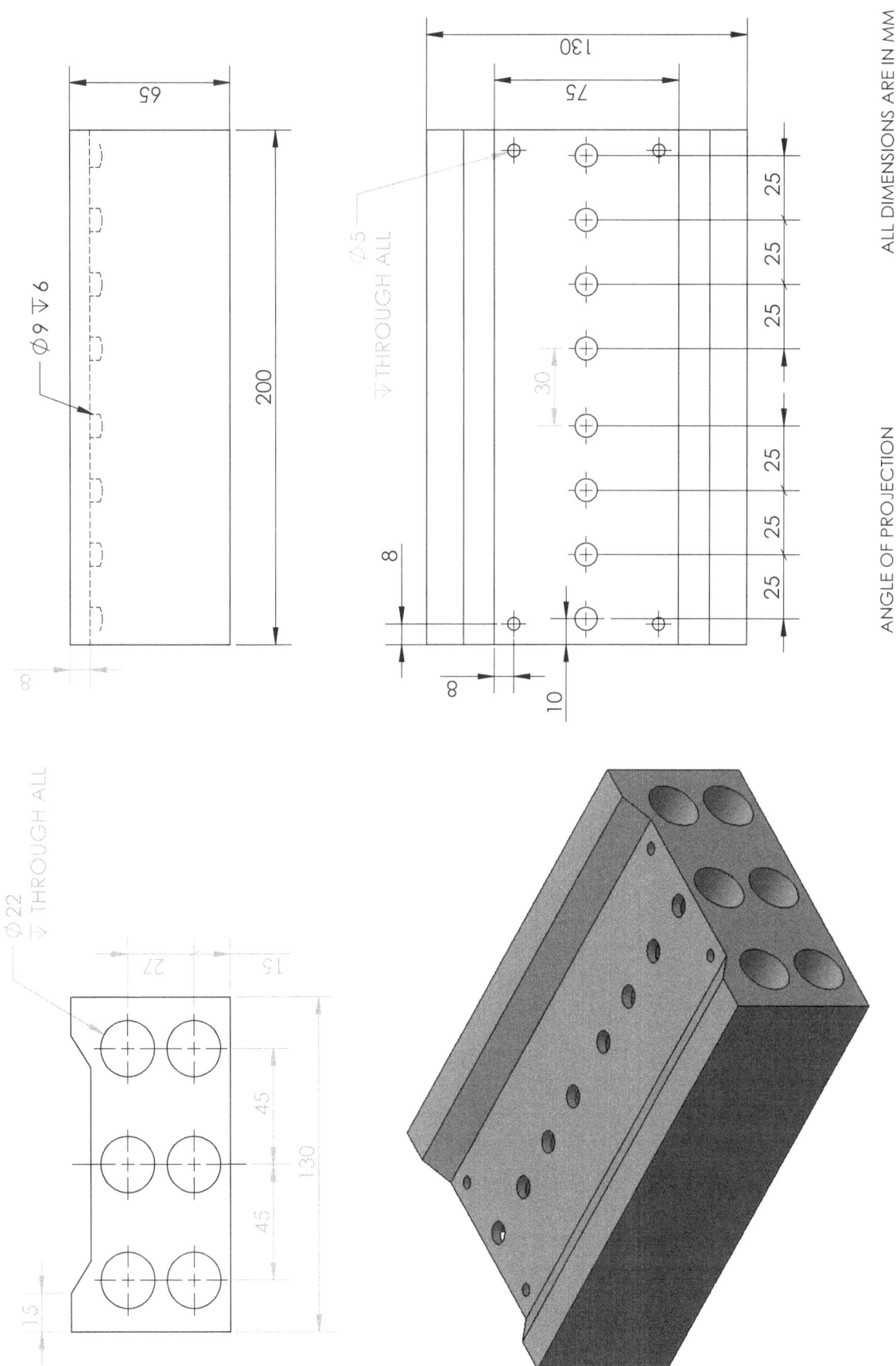

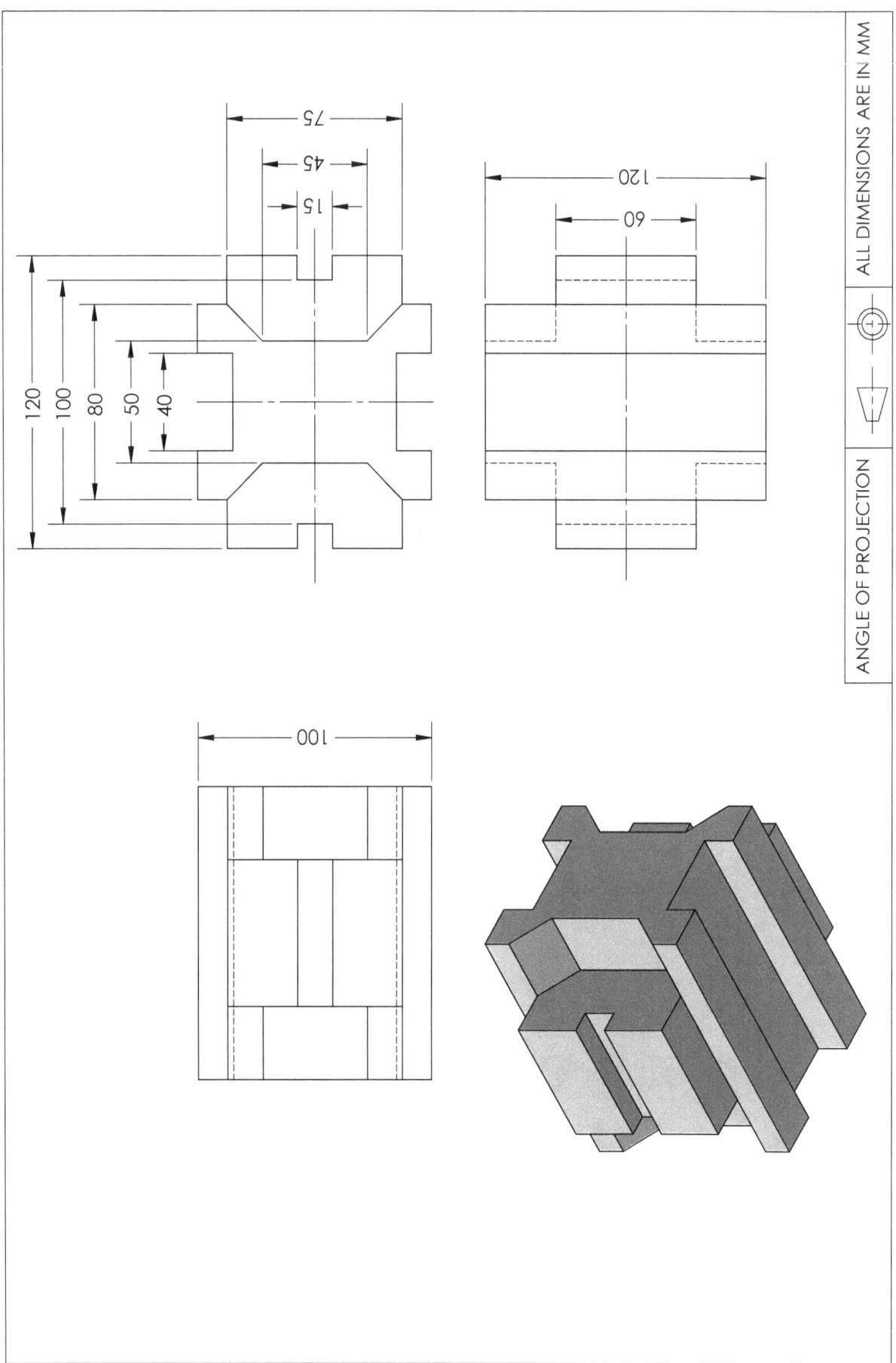

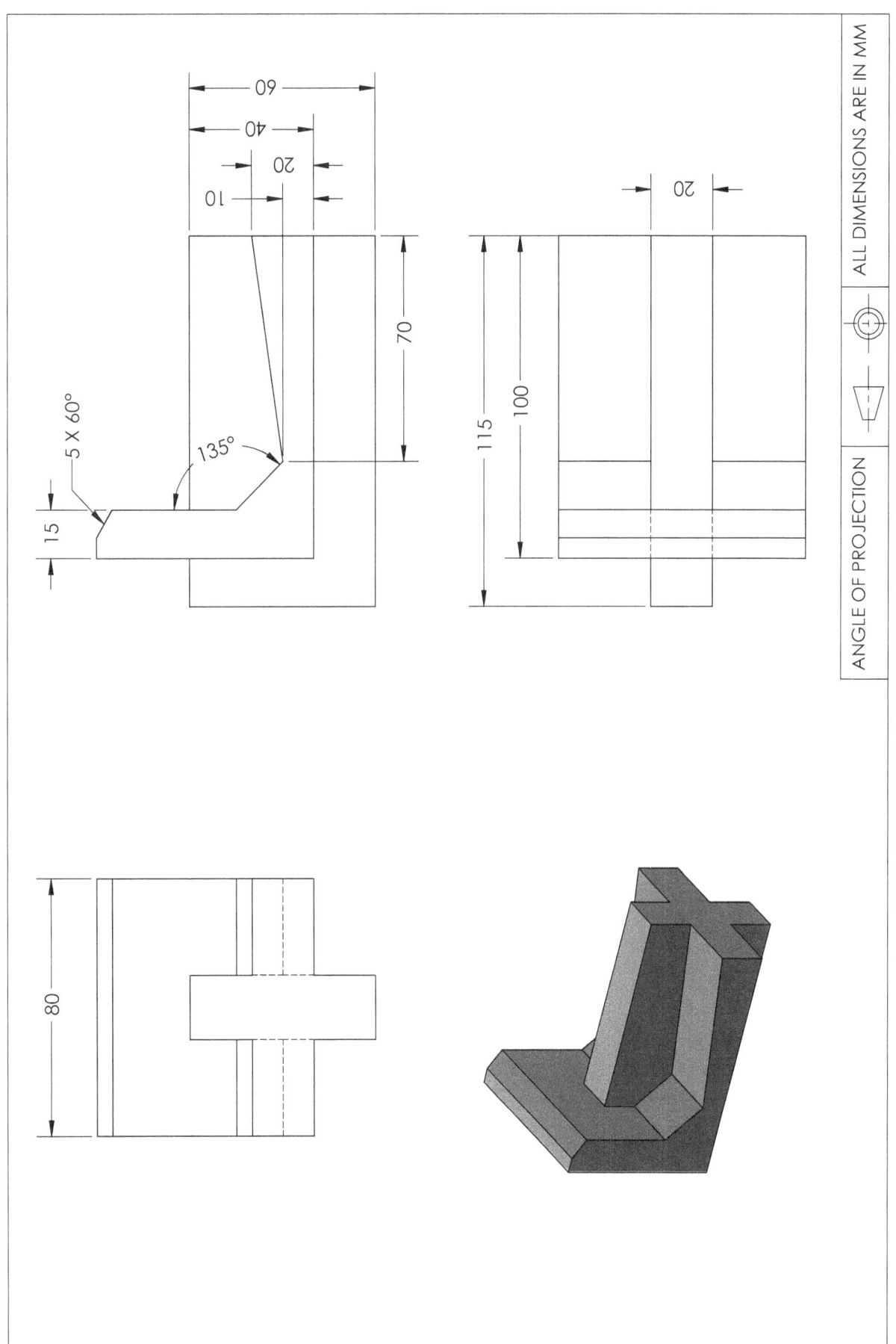

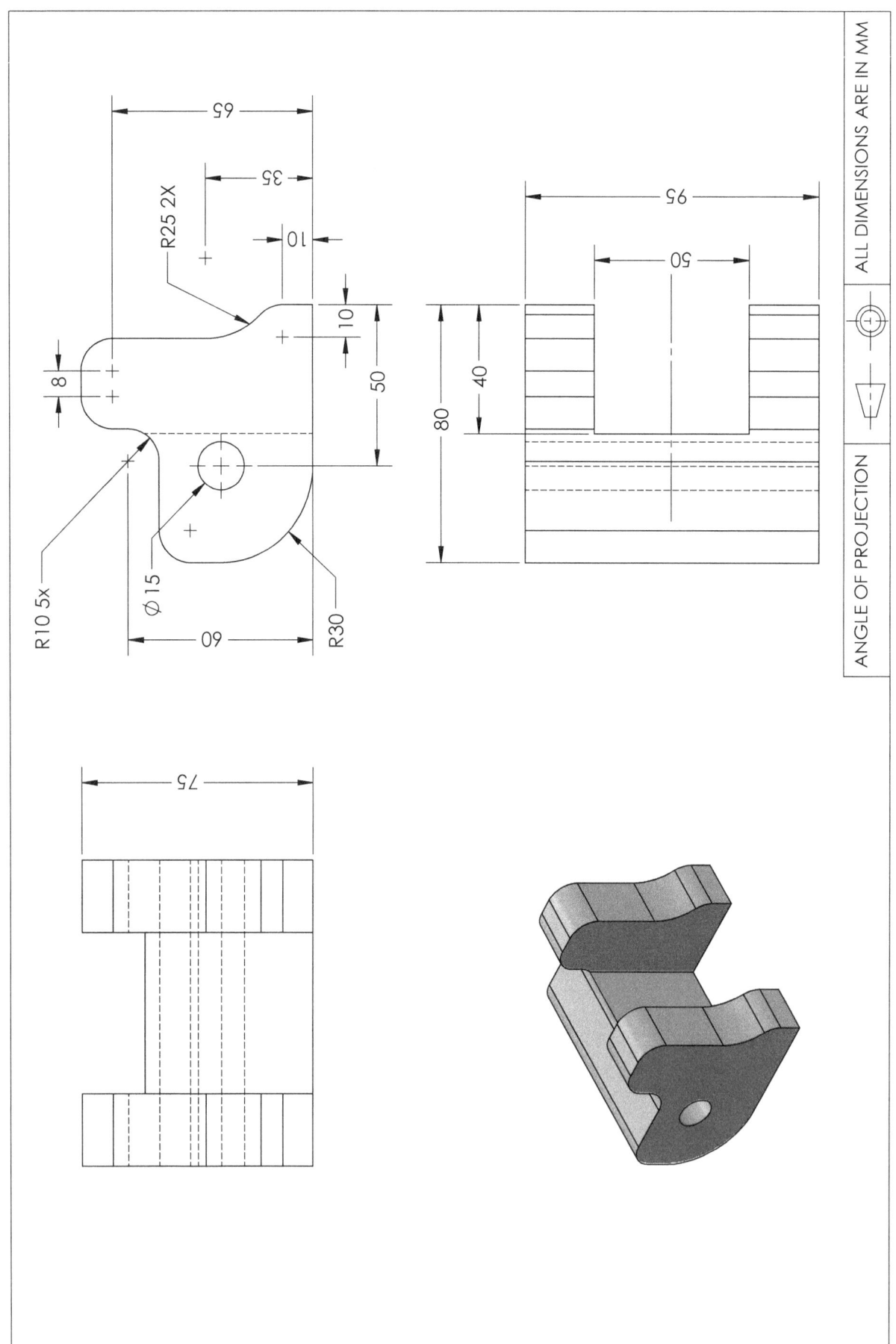

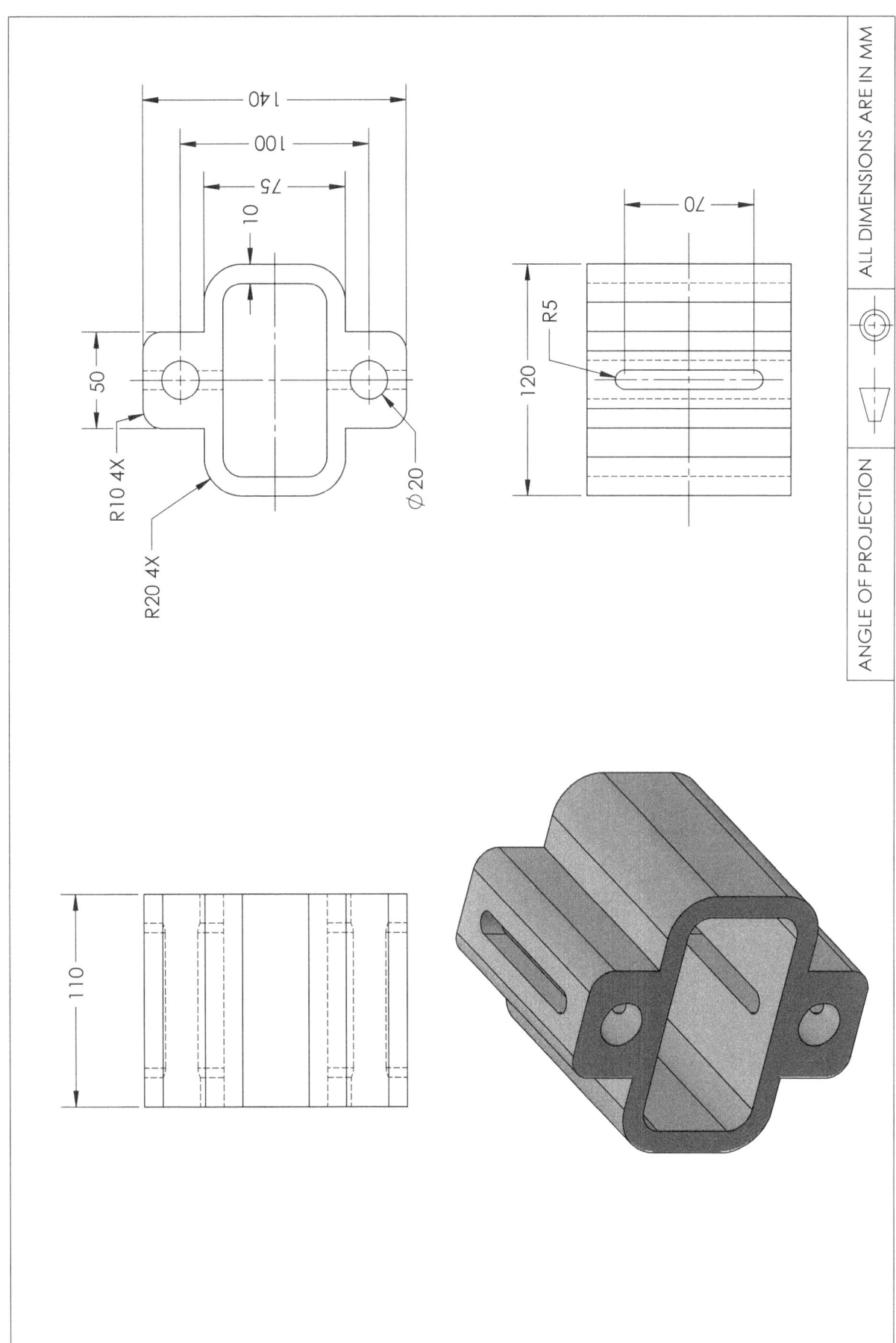

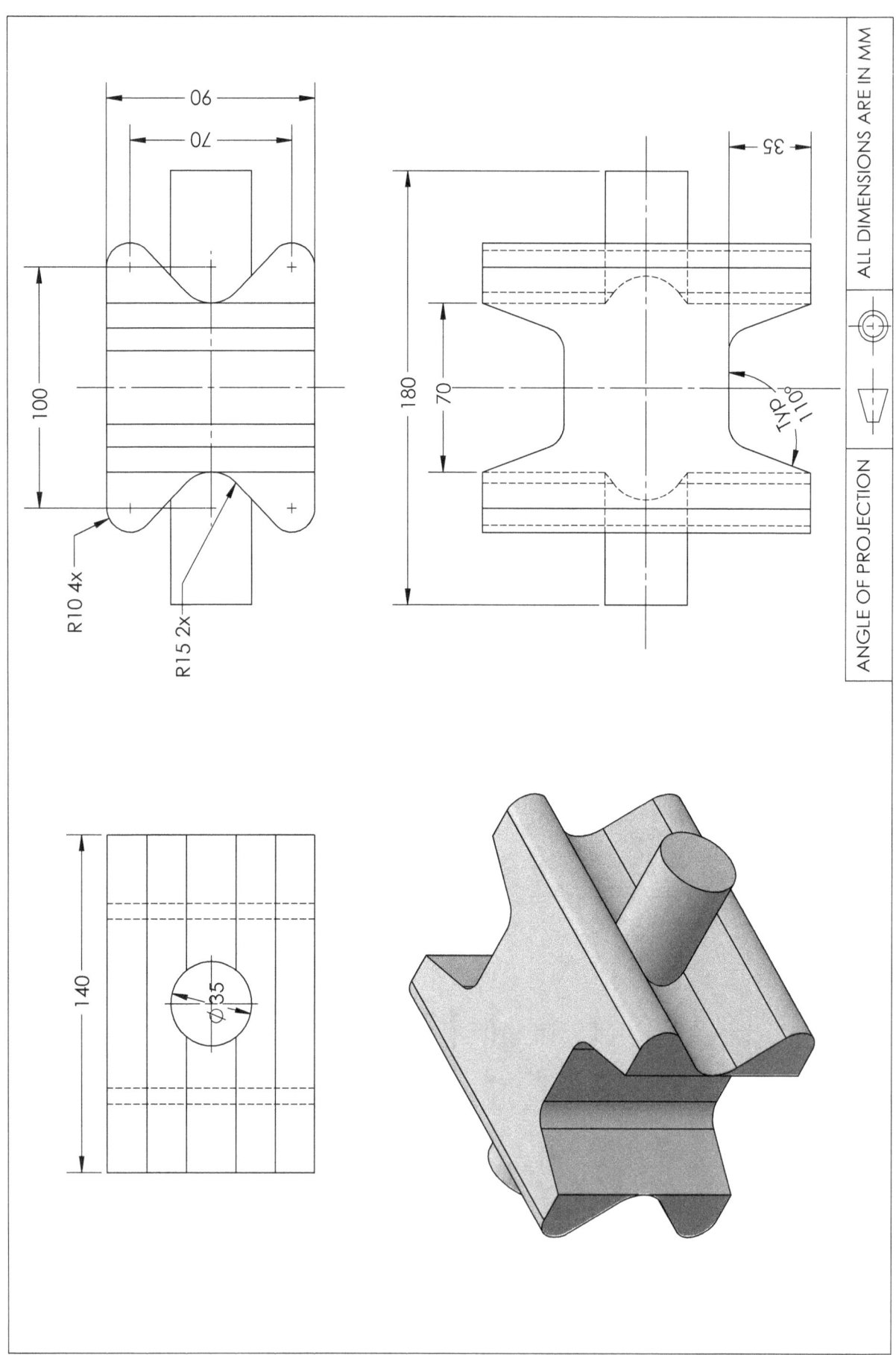

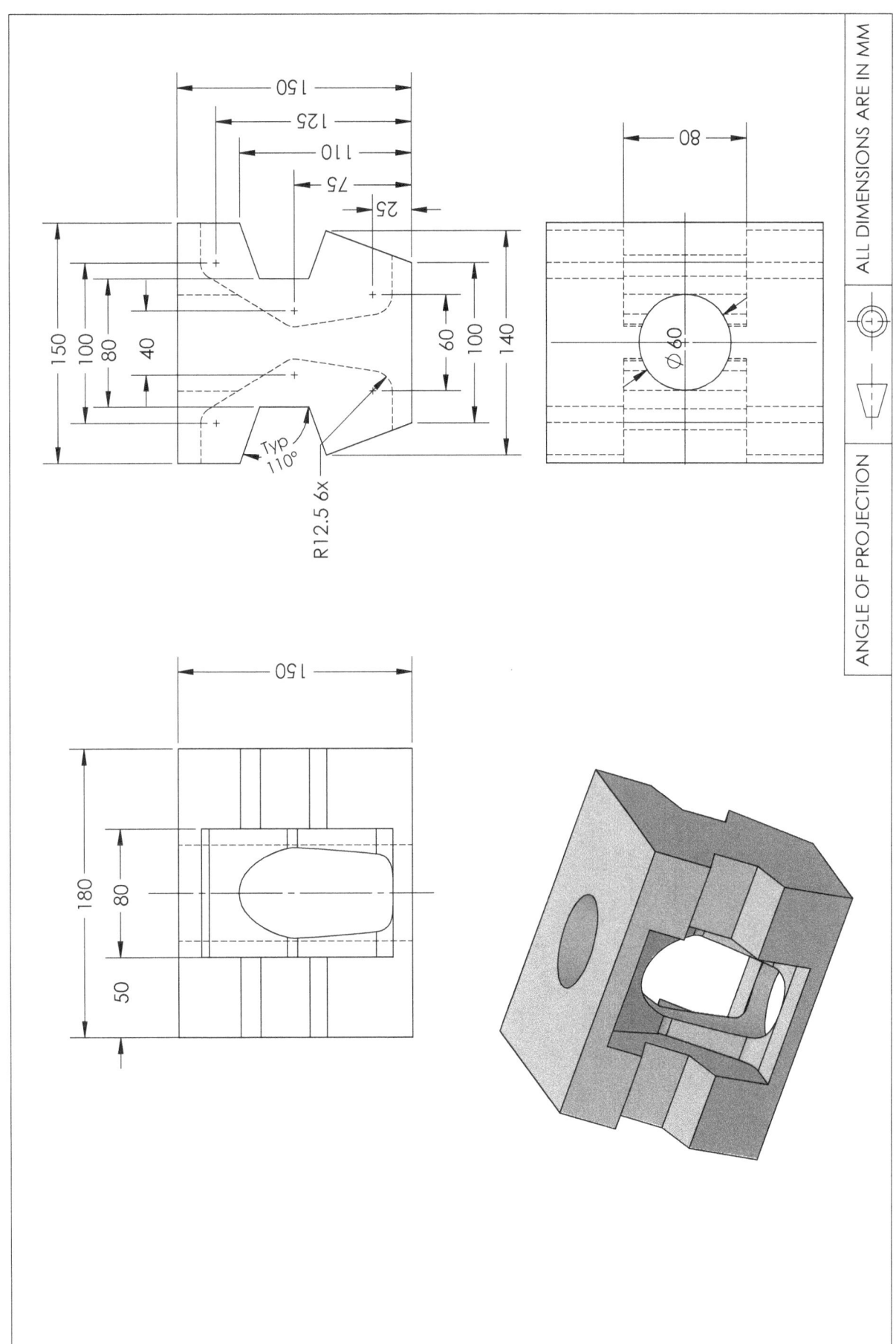

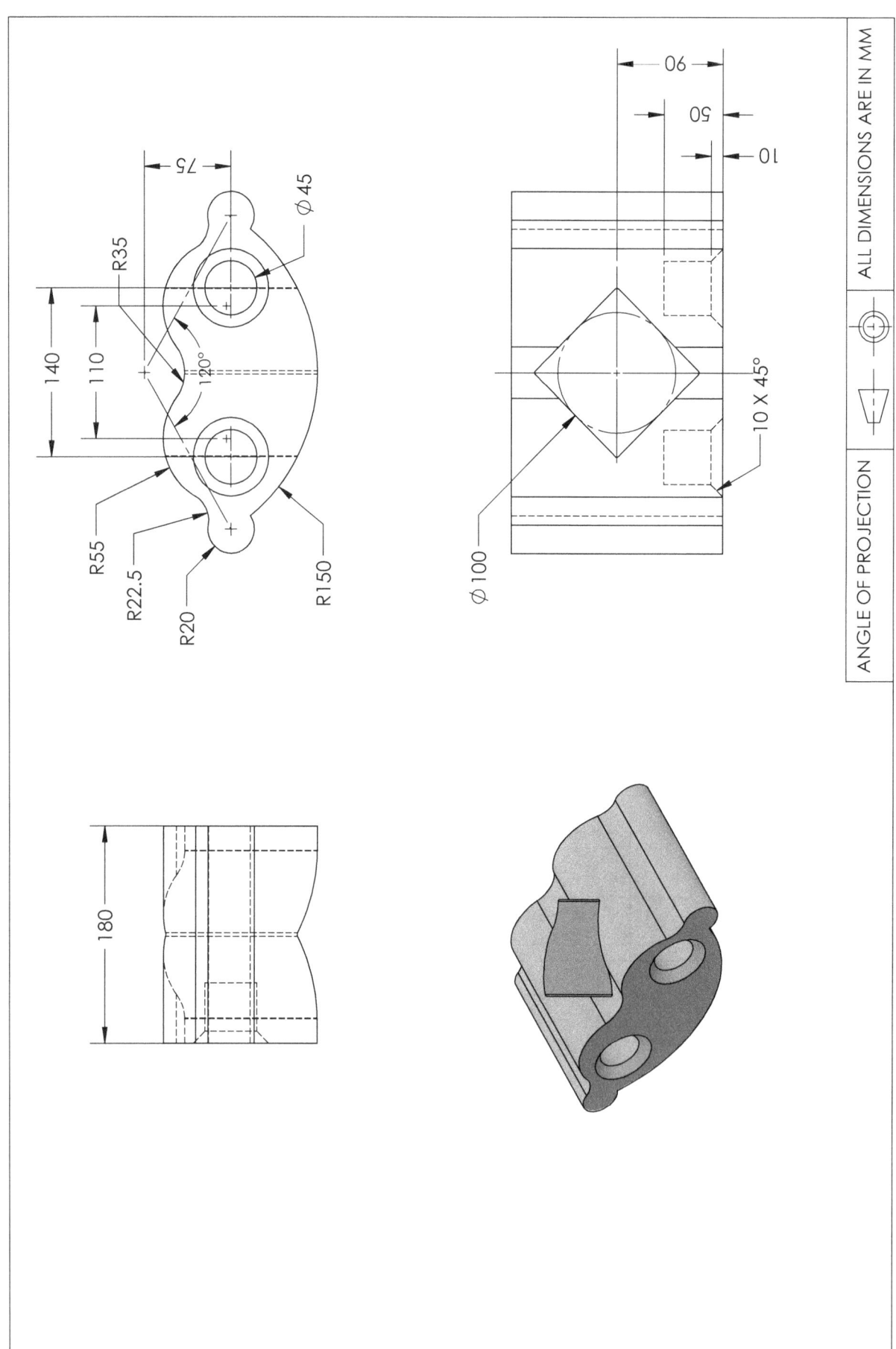

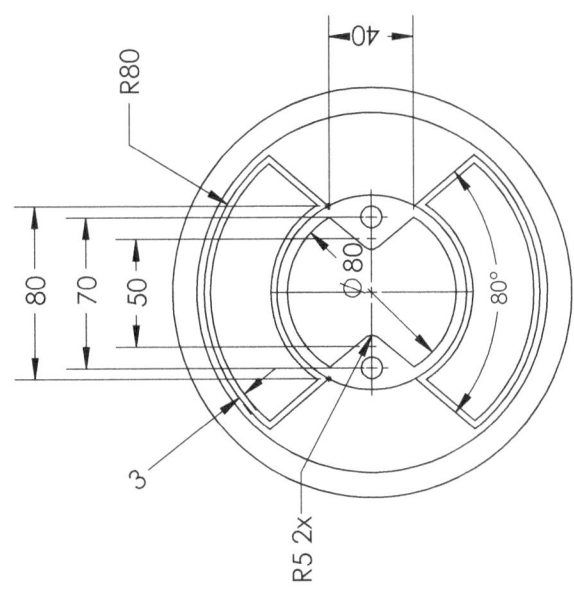

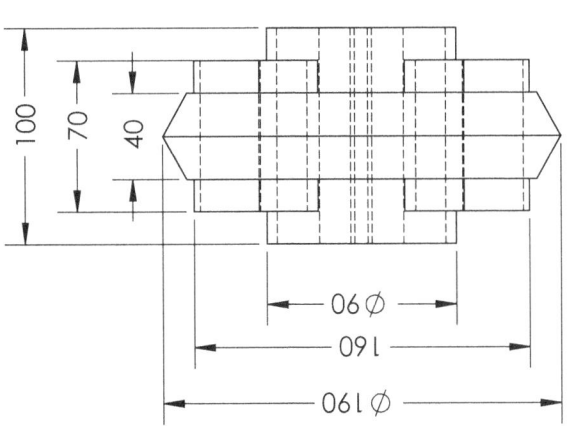
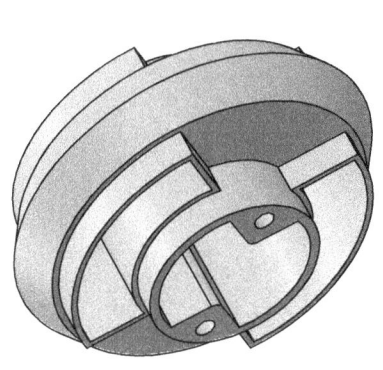

ALL DIMENSIONS ARE IN MM

ANGLE OF PROJECTION

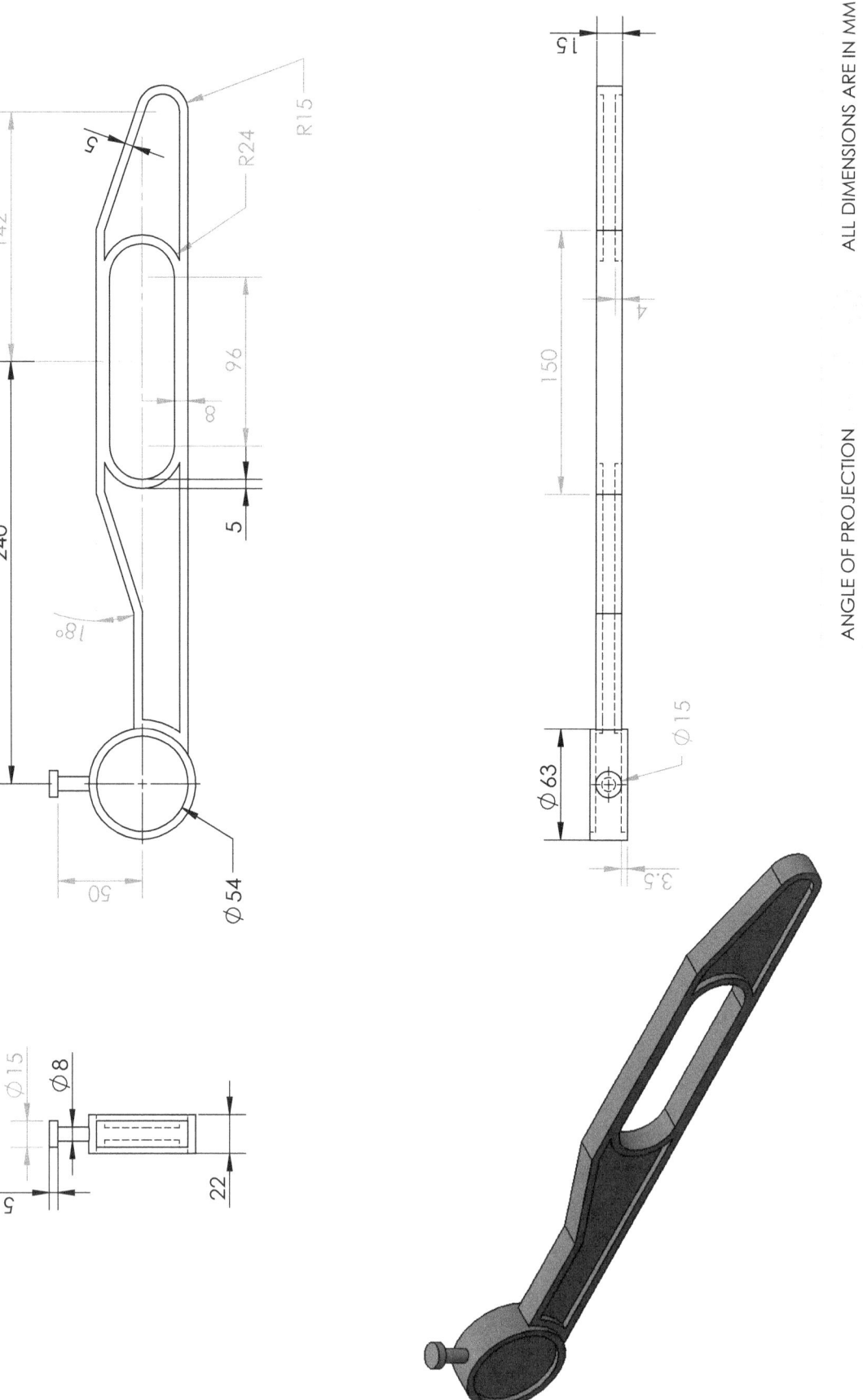

ALL DIMENSIONS ARE IN MM

ANGLE OF PROJECTION

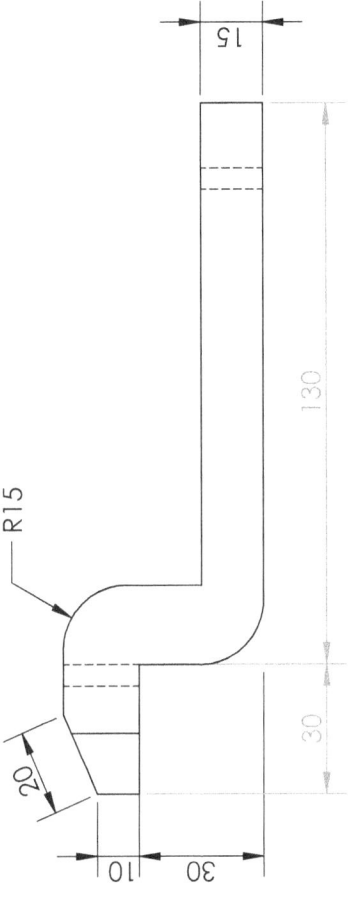
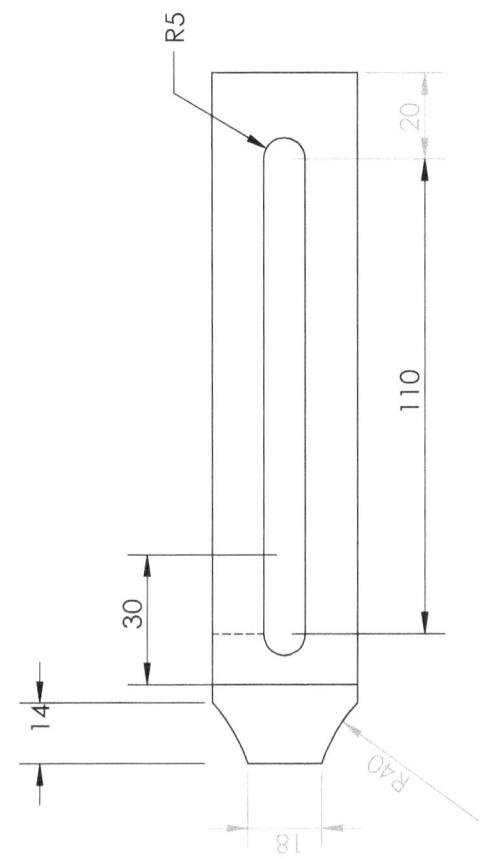
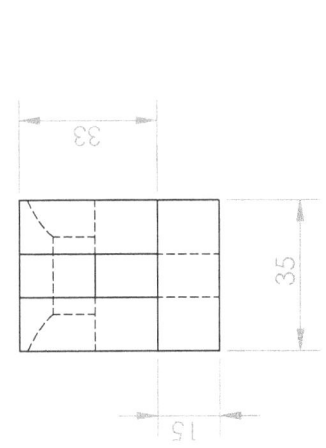
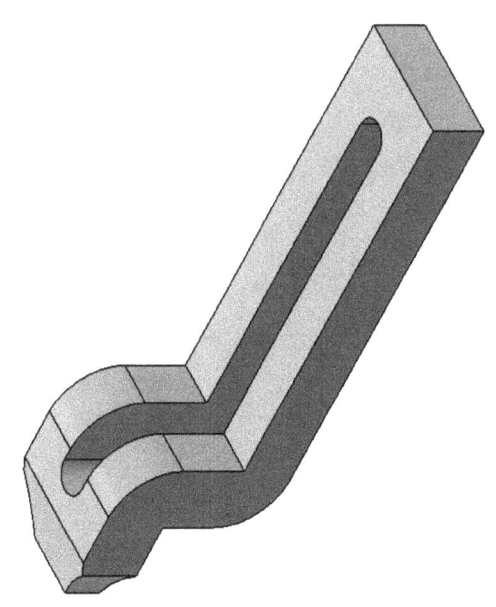

ALL DIMENSIONS ARE IN MM

ANGLE OF PROJECTION

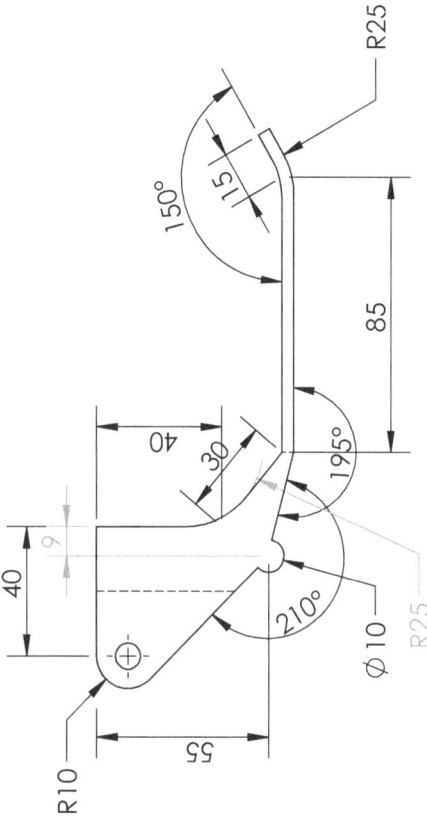
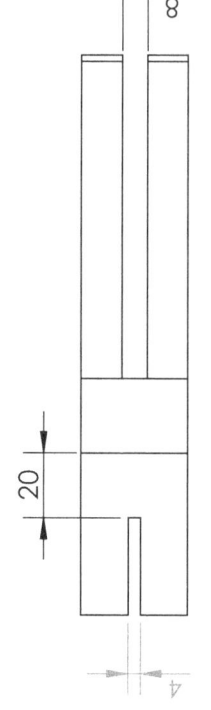
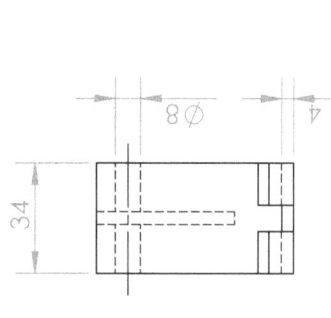

ALL DIMENSIONS ARE IN MM

ANGLE OF PROJECTION

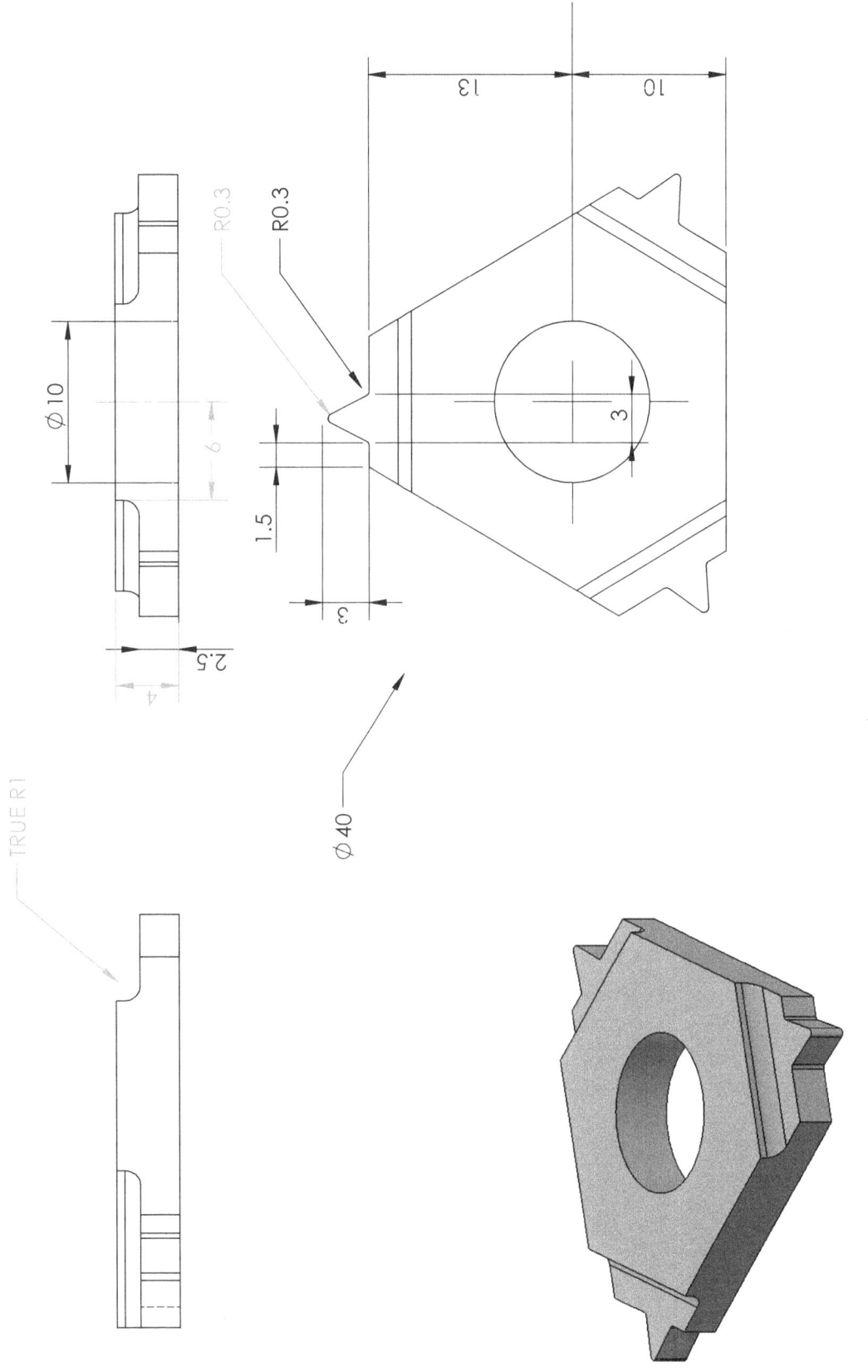

ALL DIMENSIONS ARE IN MM

ANGLE OF PROJECTION

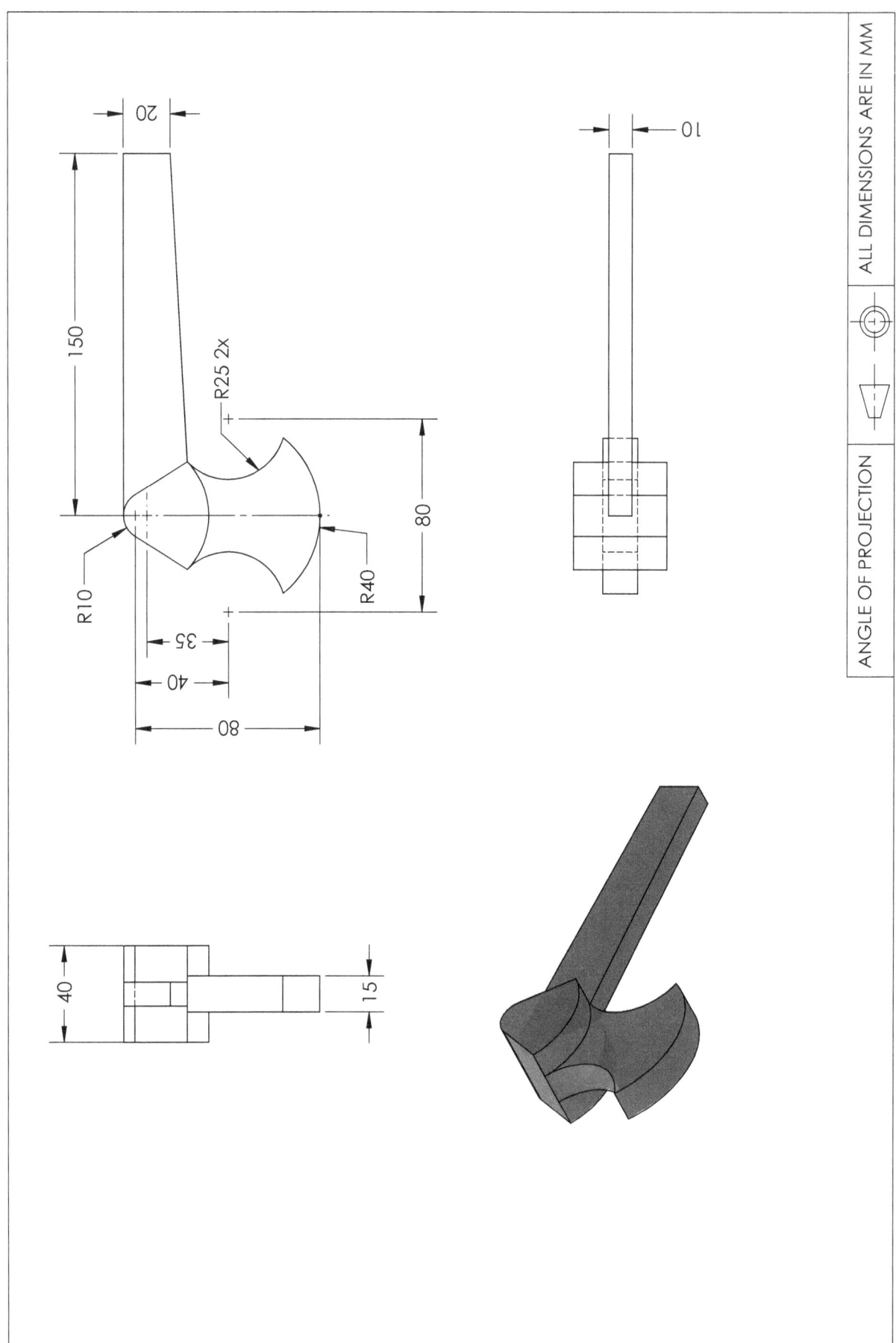

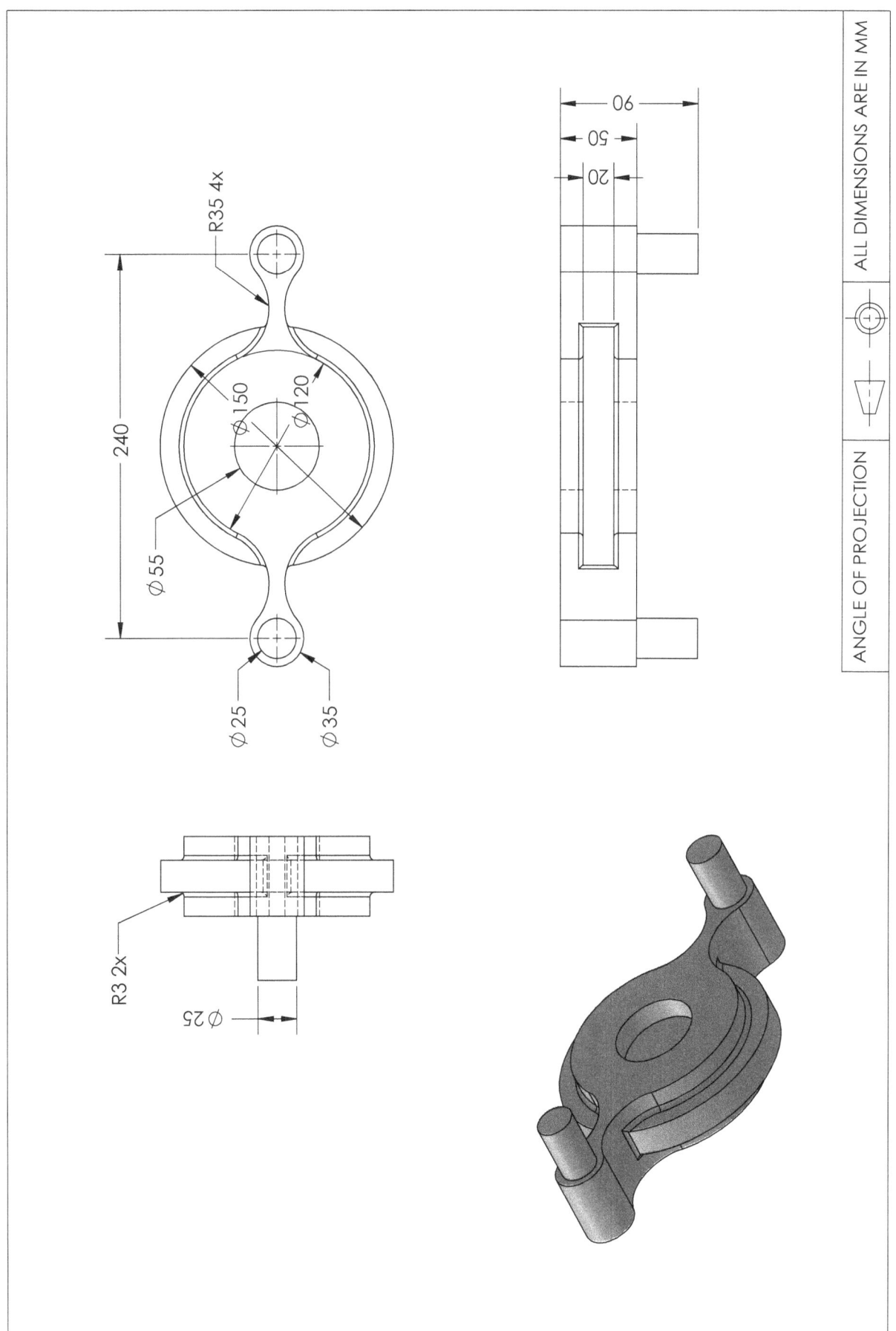

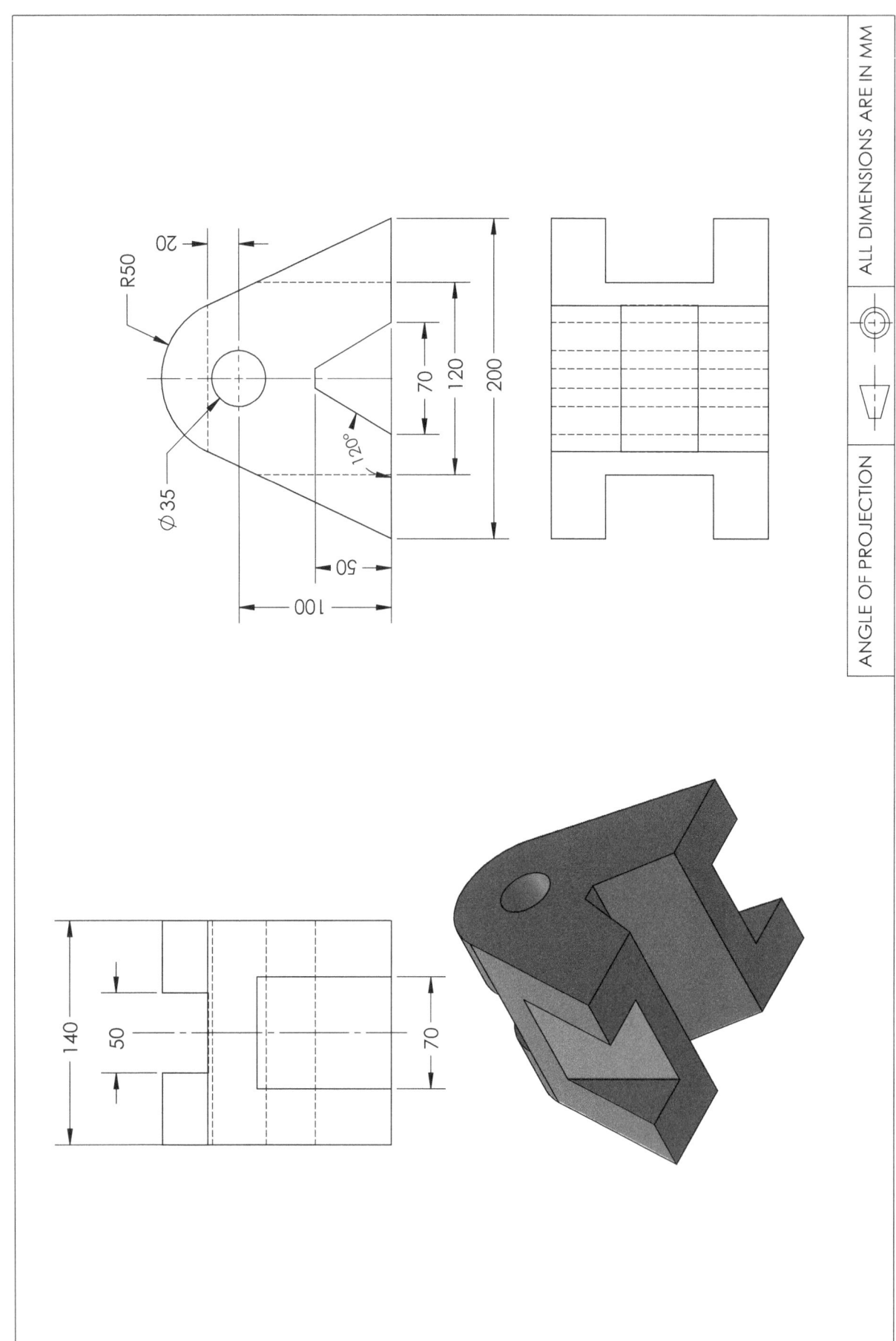

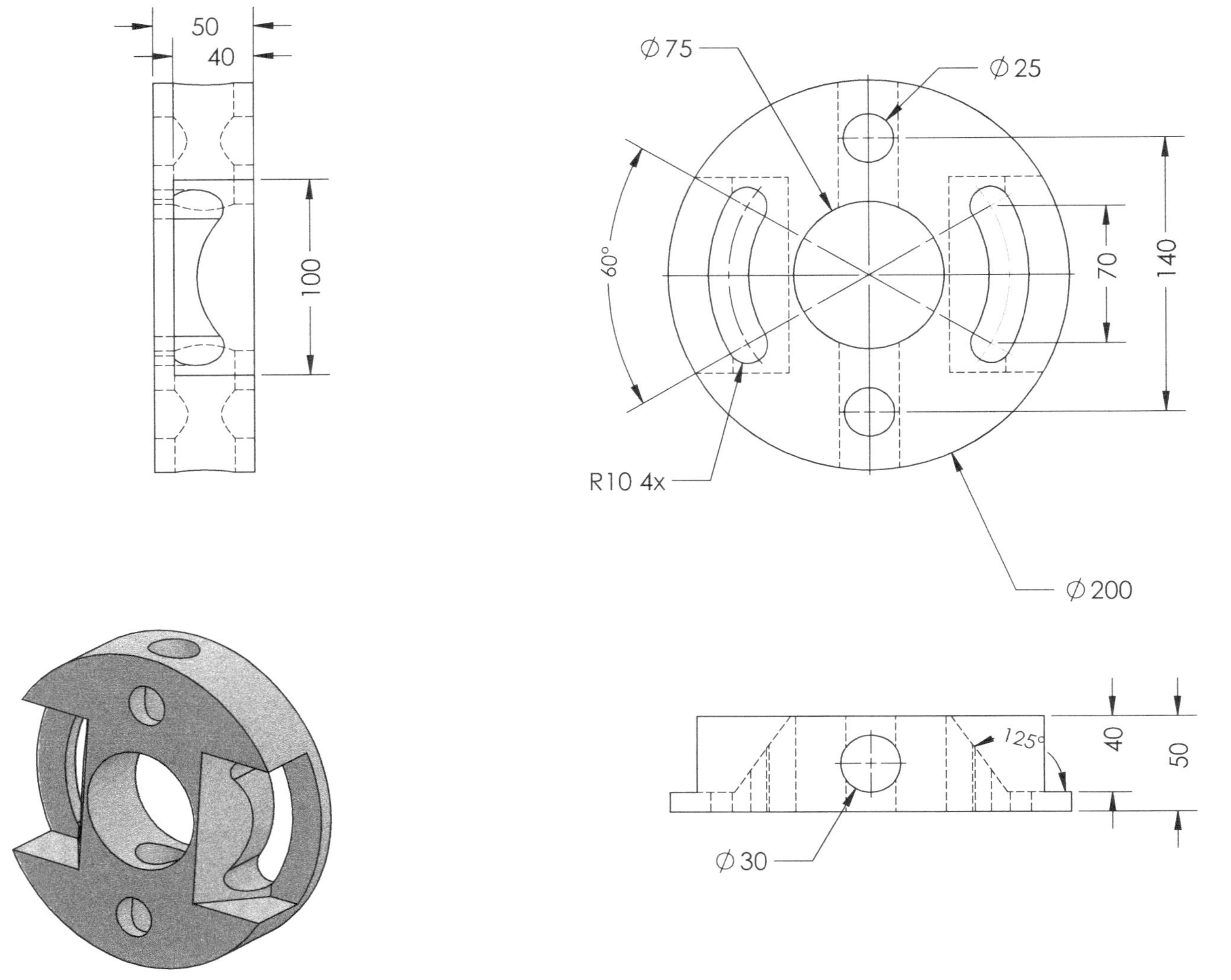

ALL DIMENSIONS ARE IN MM

ANGLE OF PROJECTION

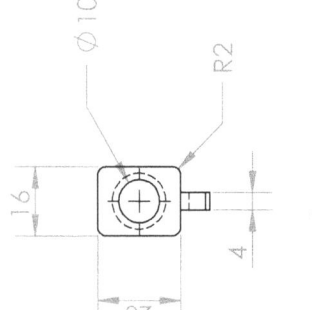
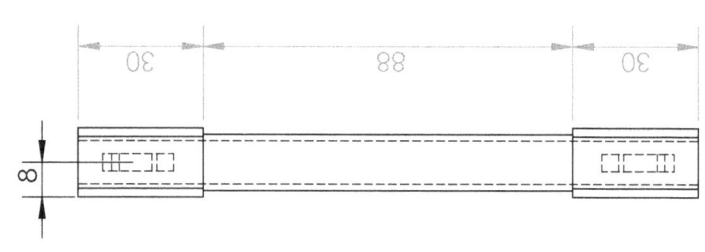
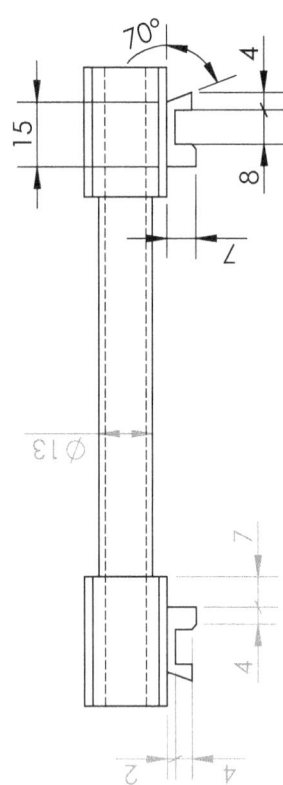
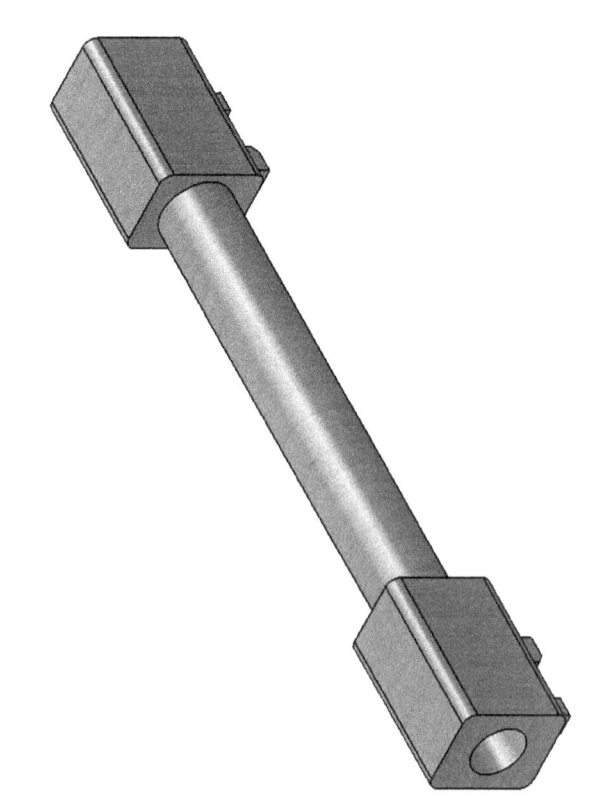

ALL DIMENSIONS ARE IN MM

ANGLE OF PROJECTION

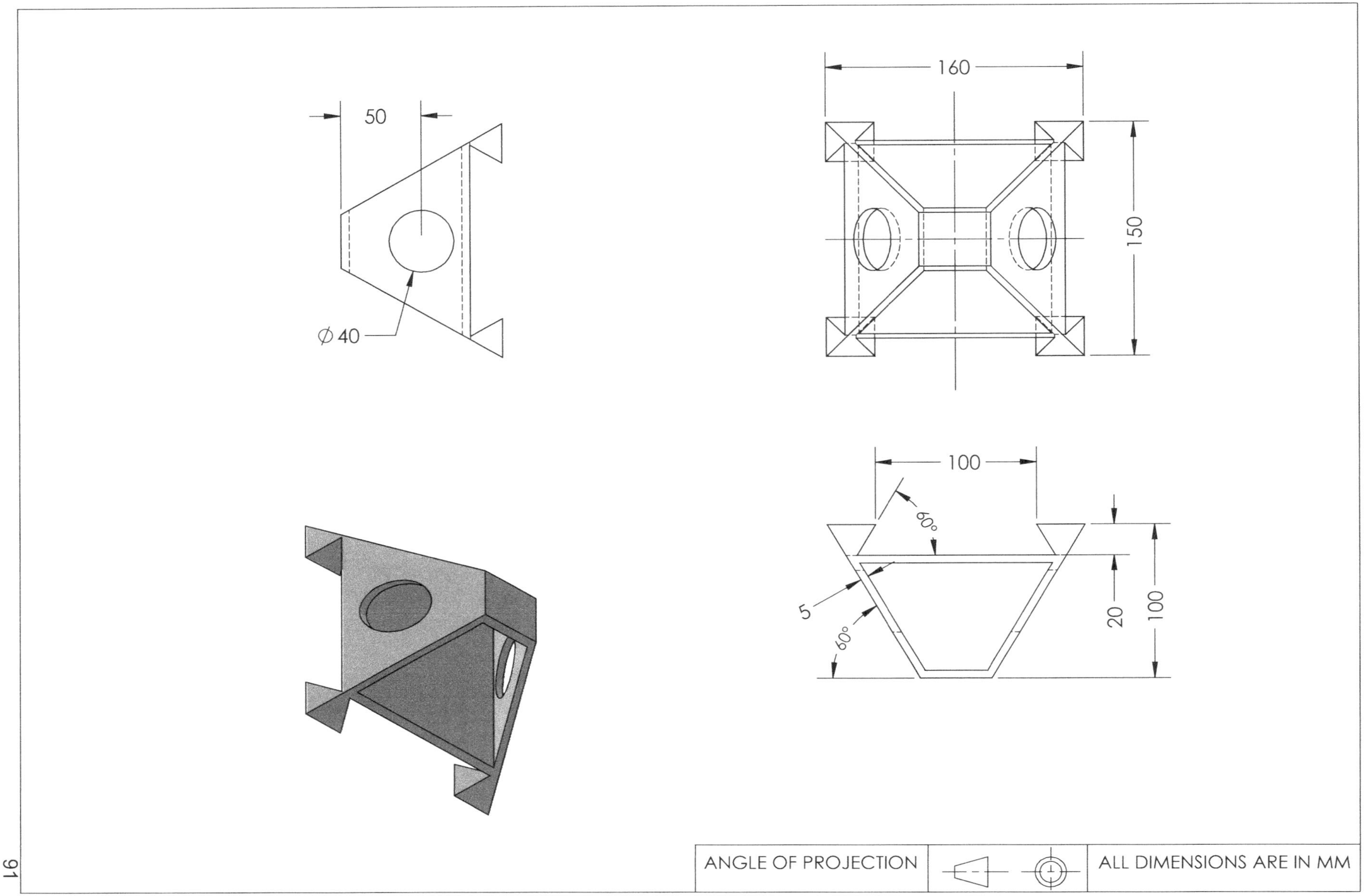

ALL DIMENSIONS ARE IN MM

ANGLE OF PROJECTION

150
160
100
20
100
60°
60°
5
50
Ø40

91

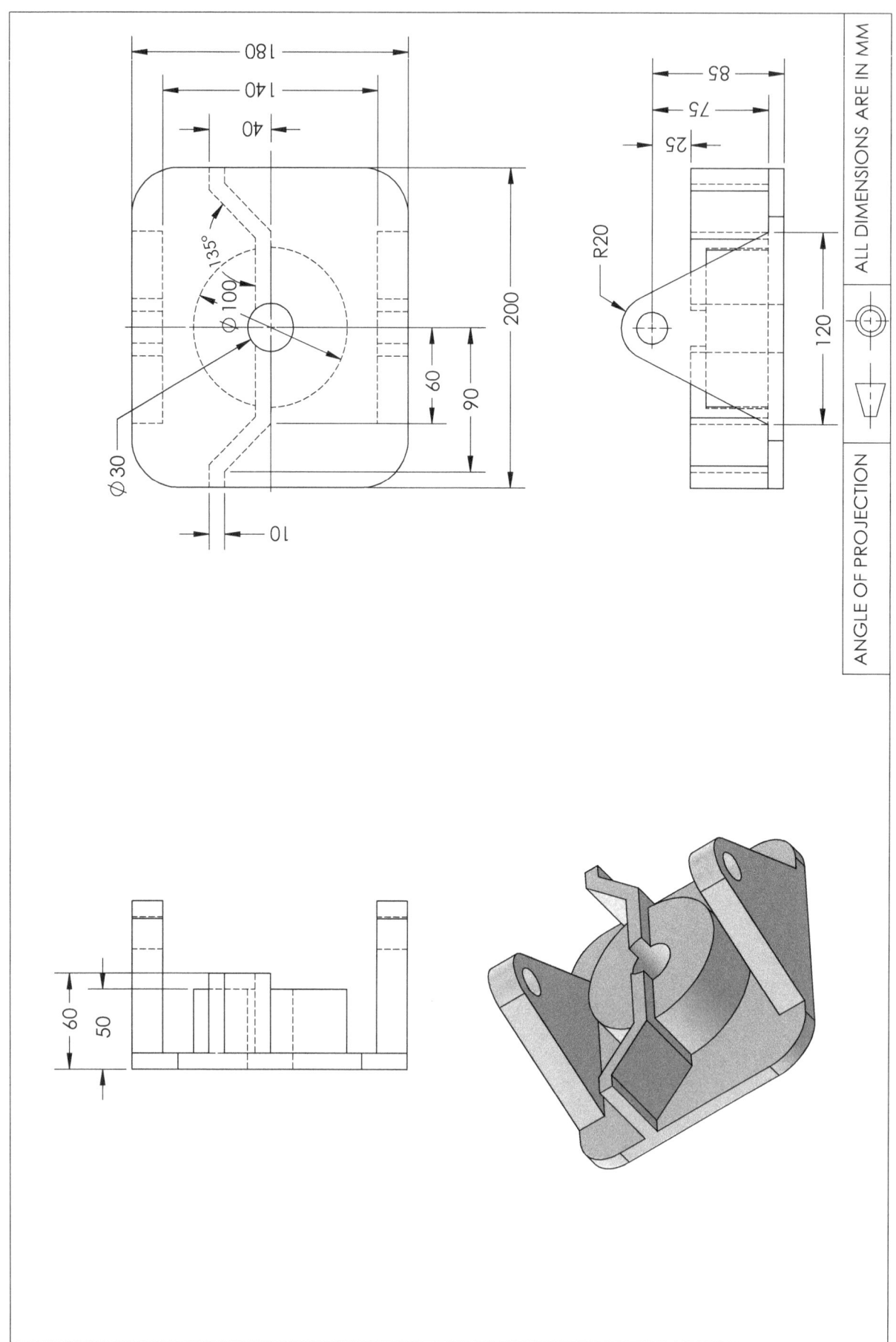

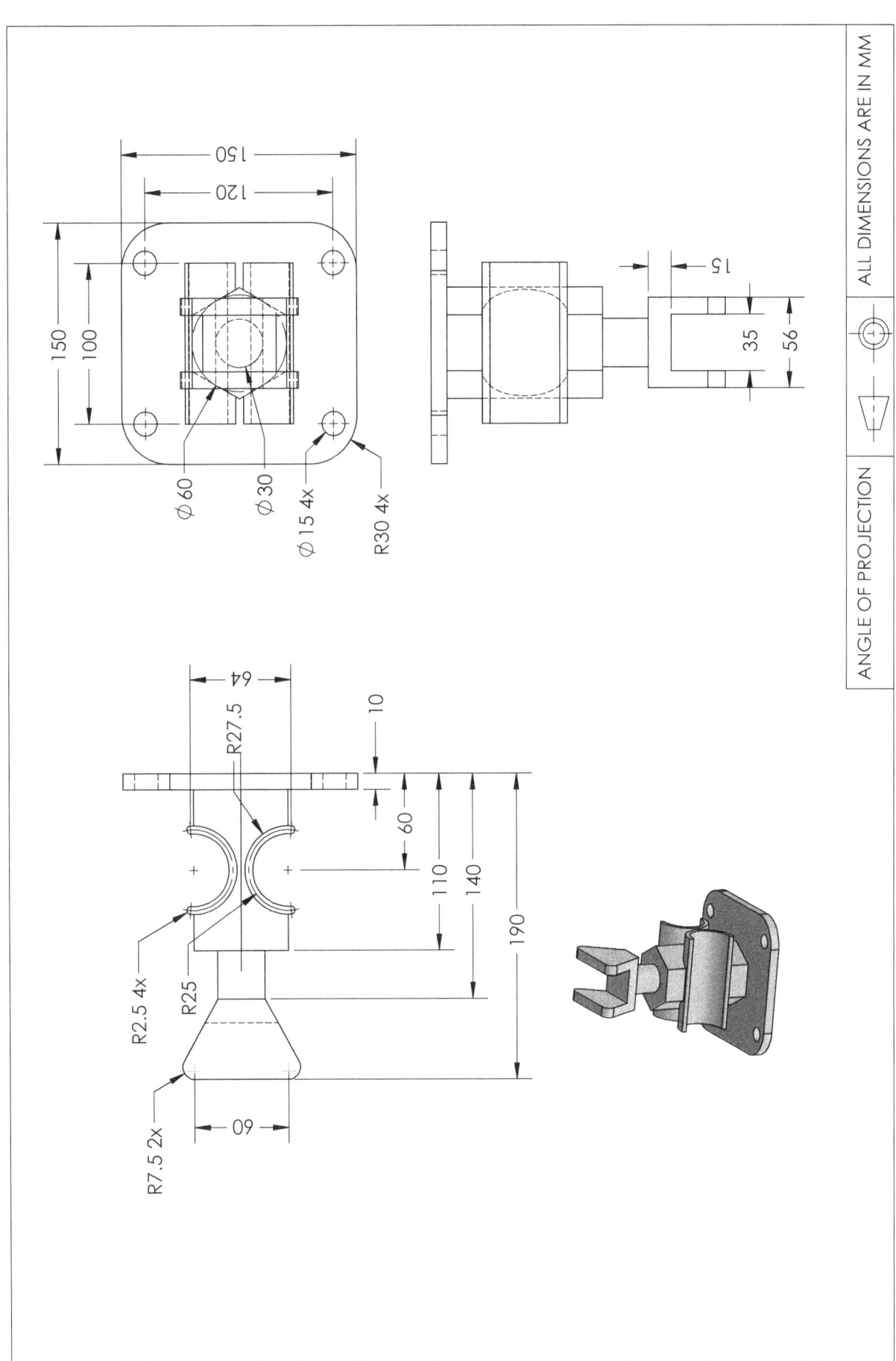

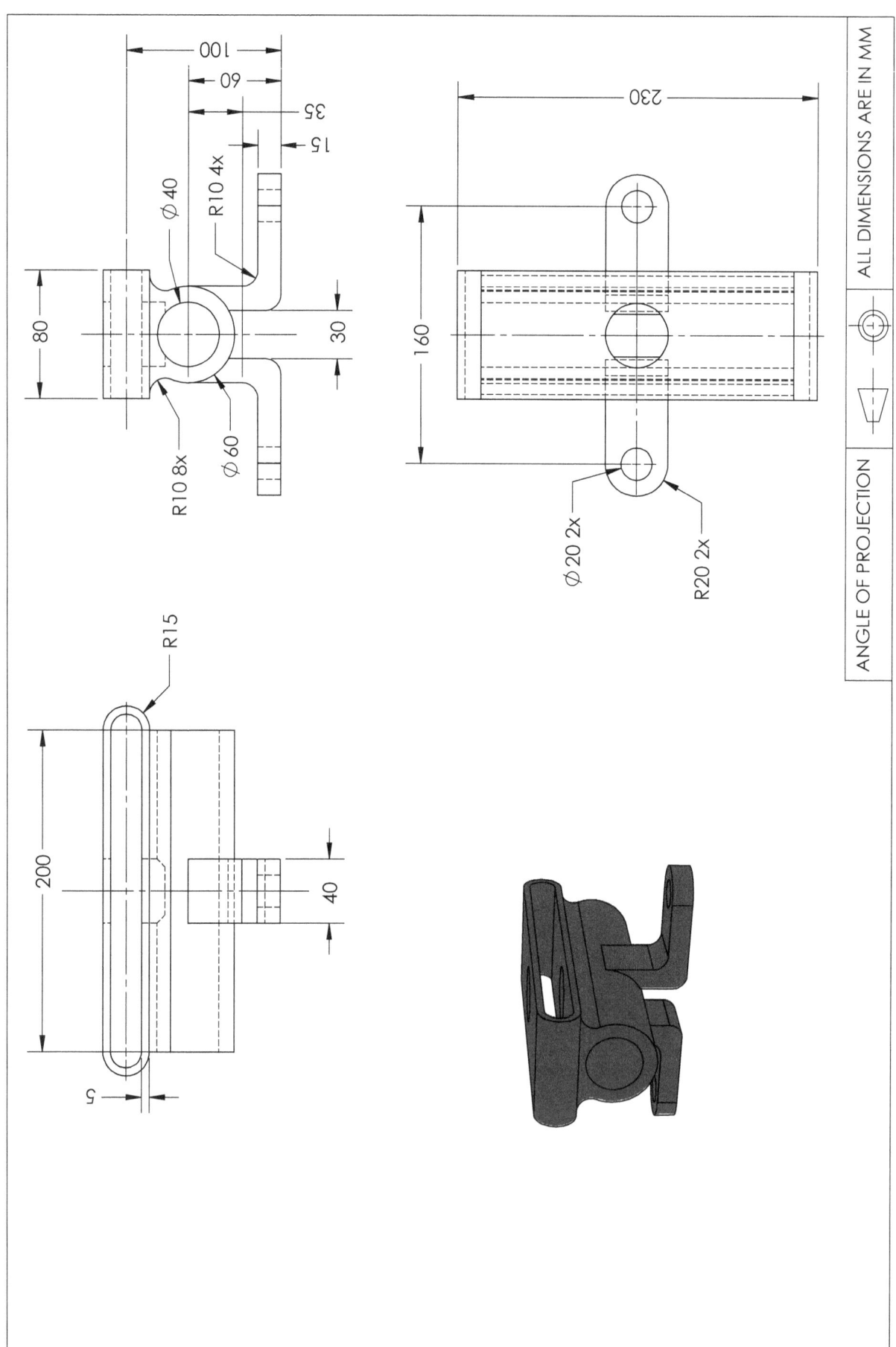

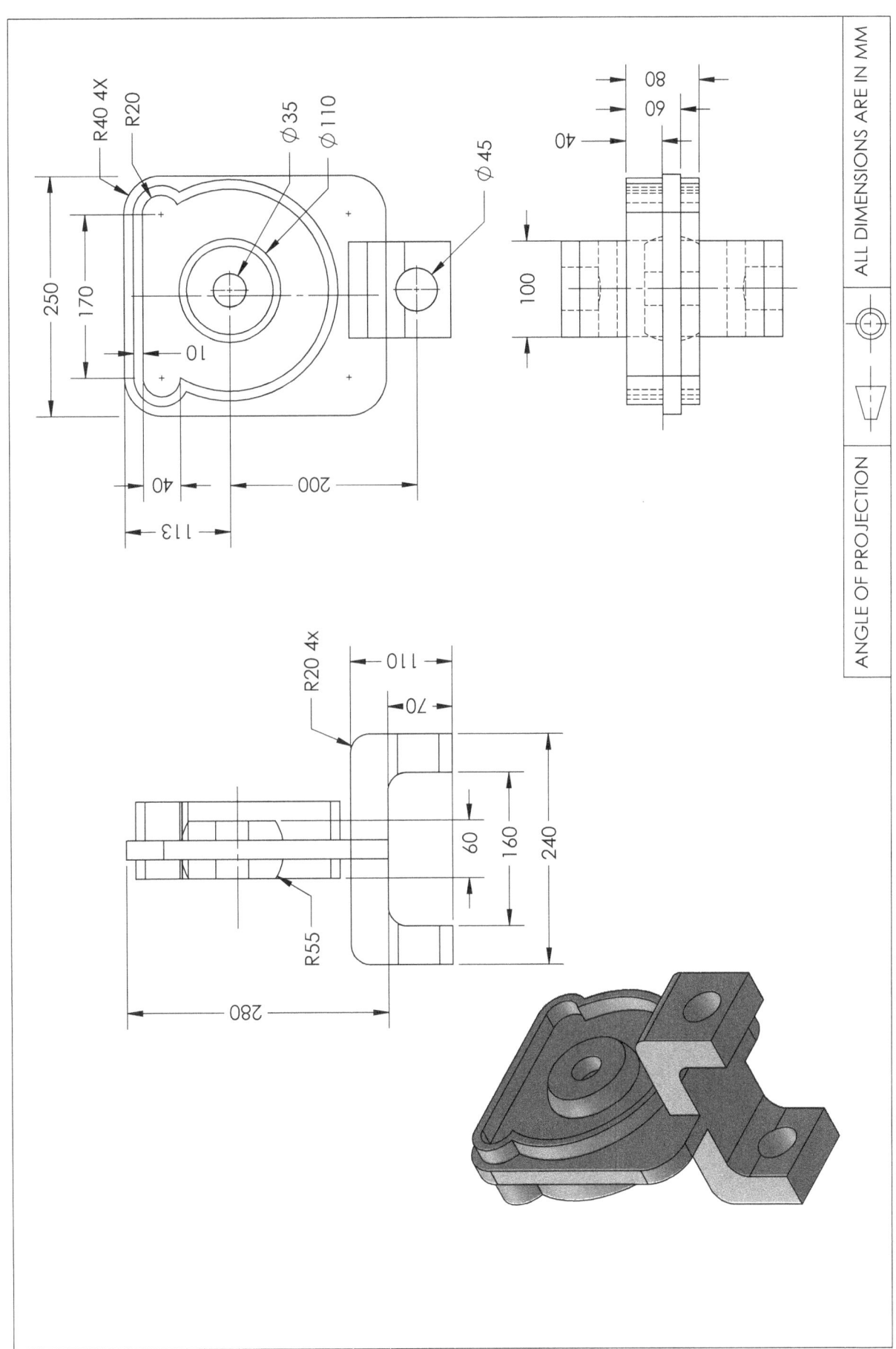

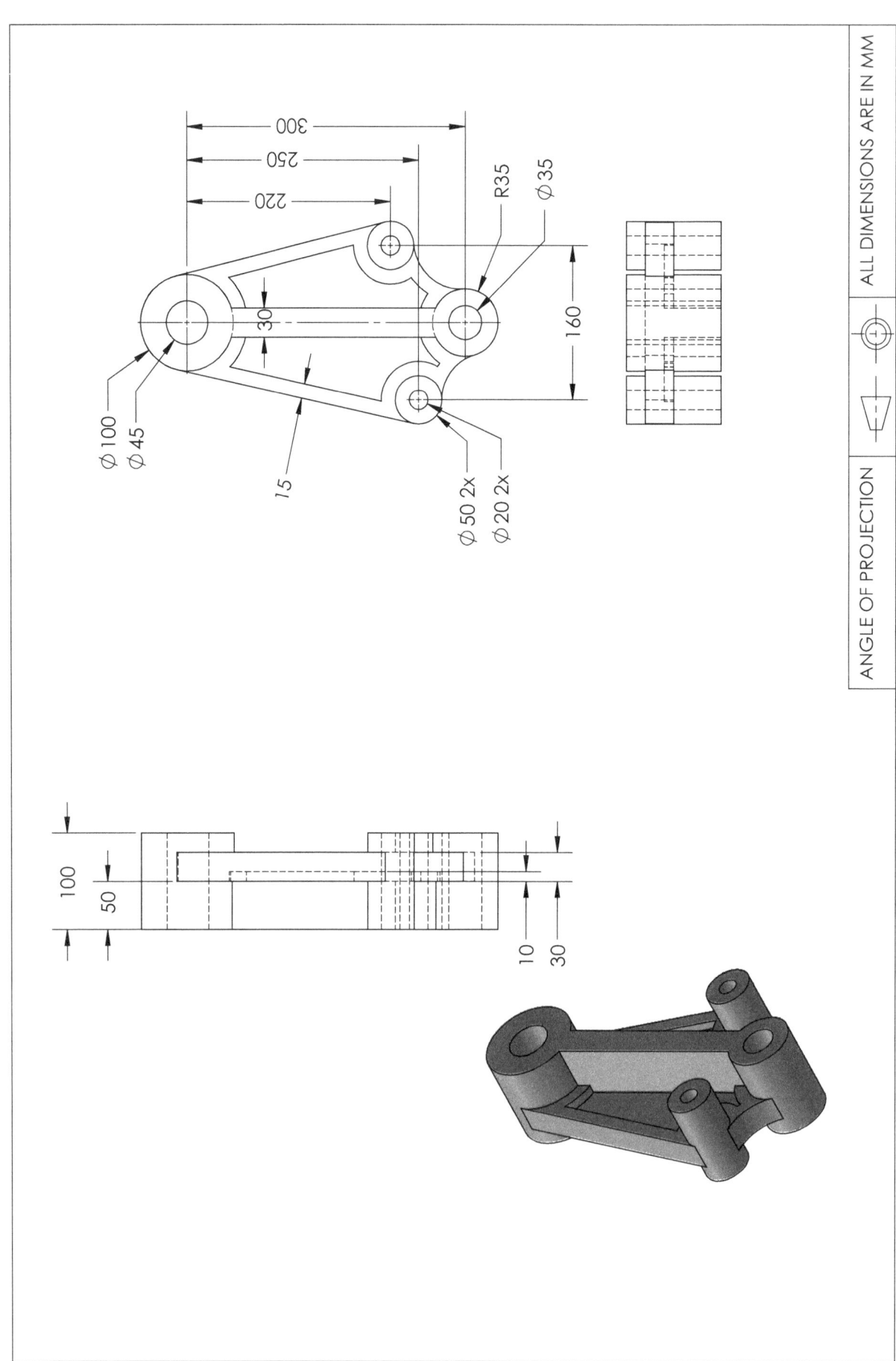

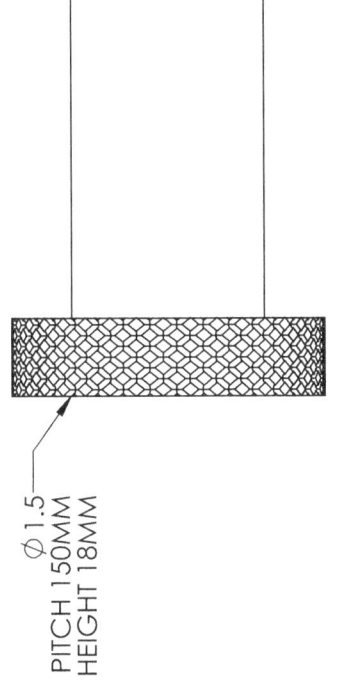
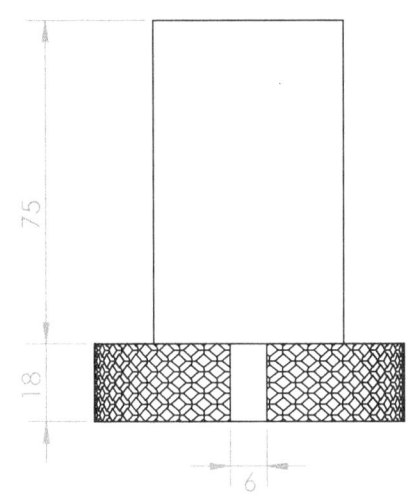

PITCH 150MM
HEIGHT 18MM
Ø1.5

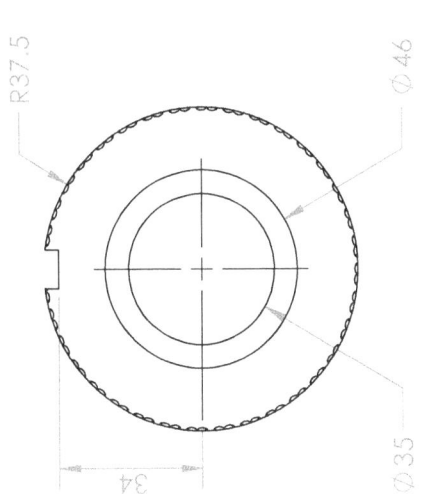
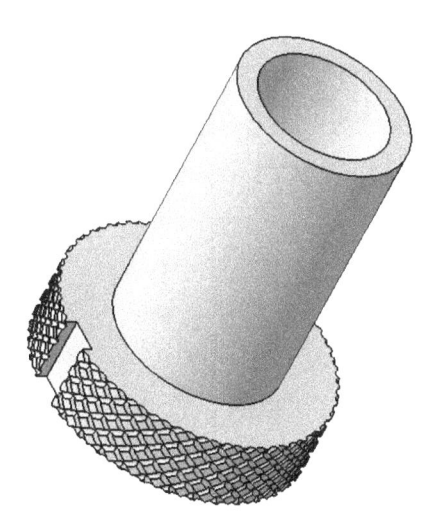

ALL DIMENSIONS ARE IN MM

ANGLE OF PROJECTION

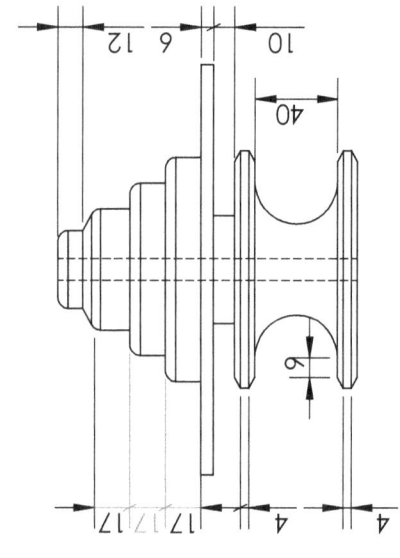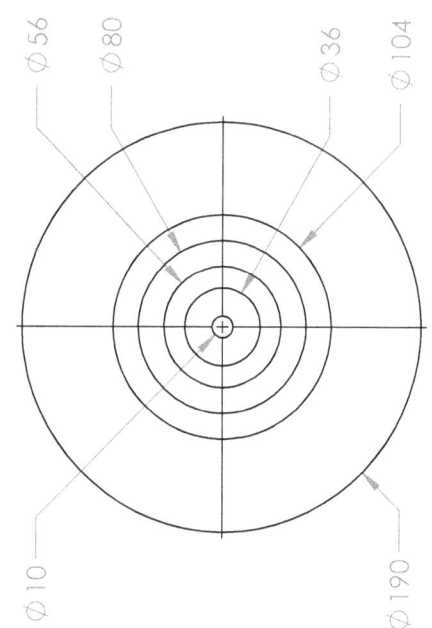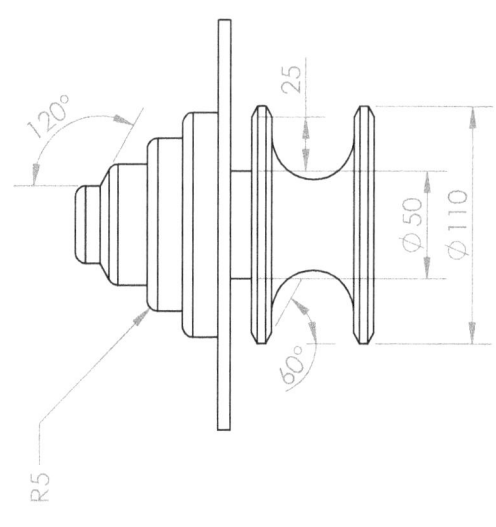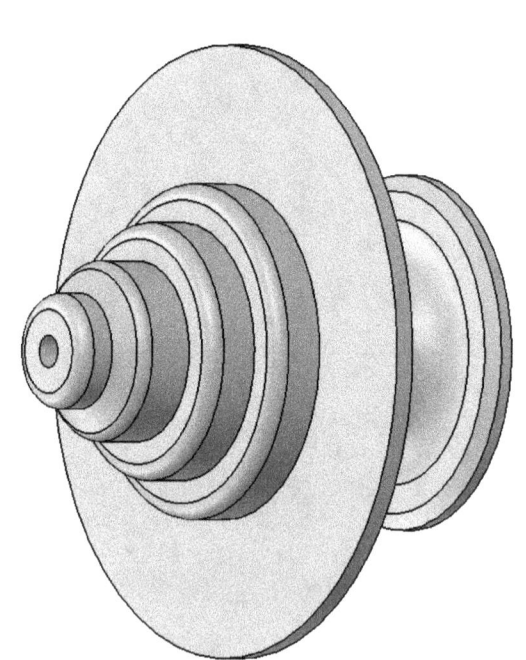

ALL DIMENSIONS ARE IN MM

ANGLE OF PROJECTION

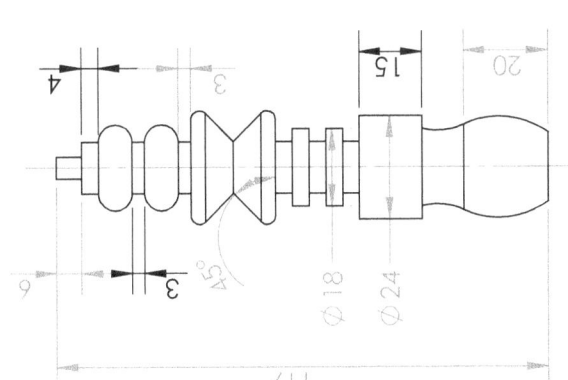

SQUA 5

ALL DIMENSIONS ARE IN MM

ANGLE OF PROJECTION

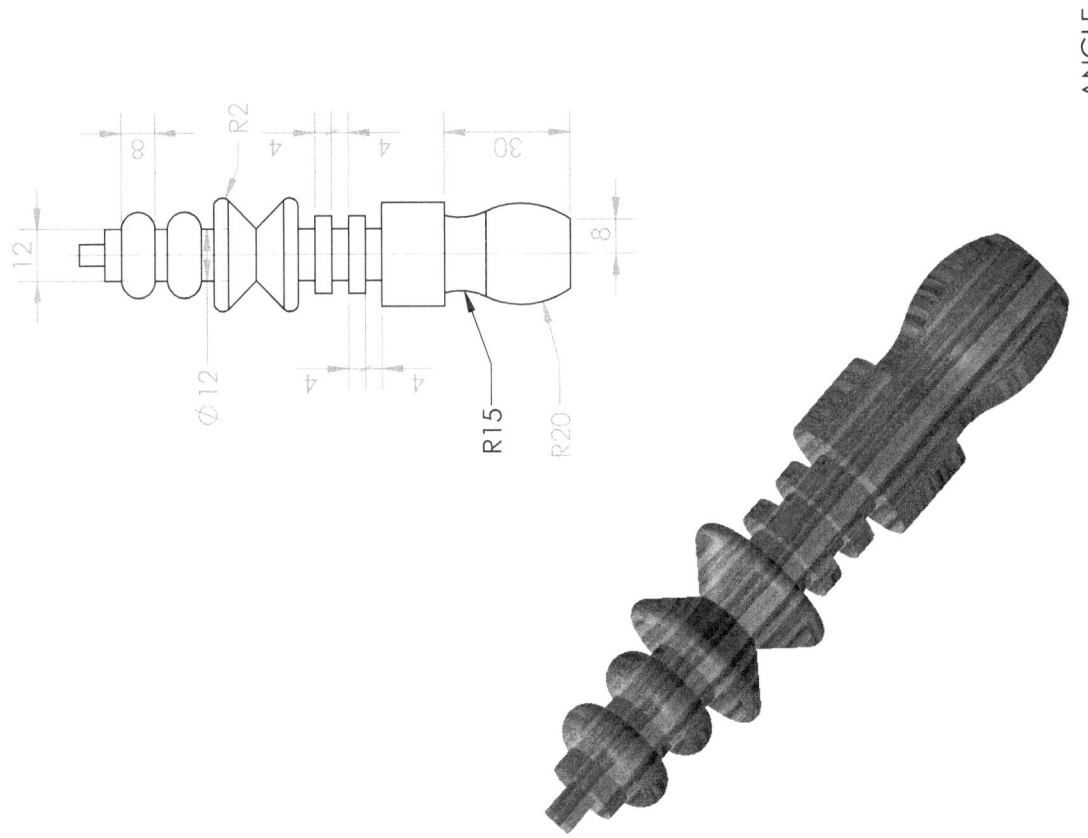

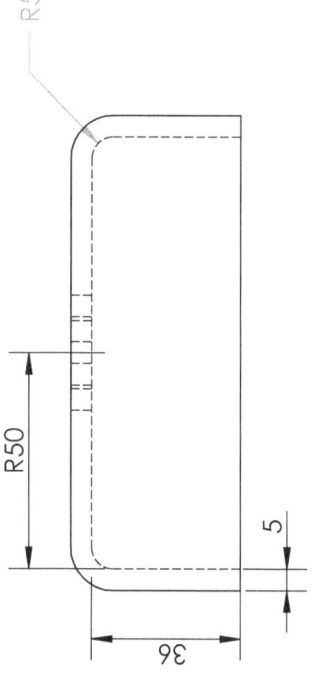
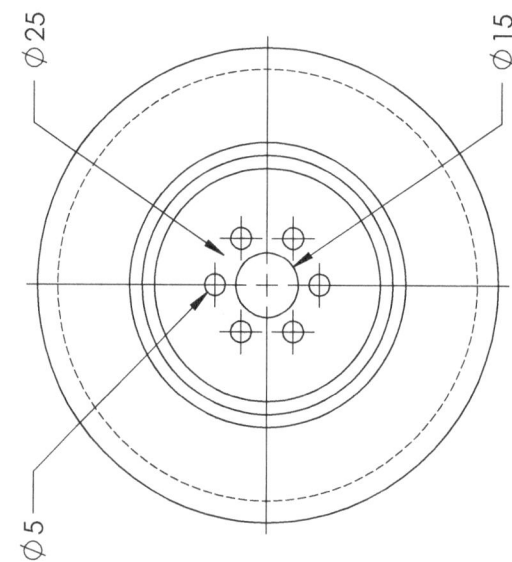
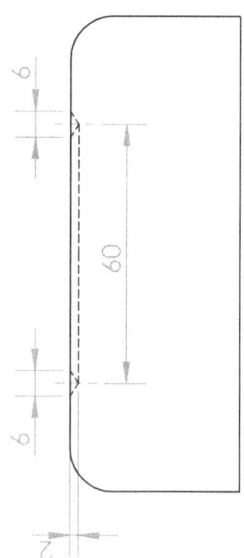

ALL DIMENSIONS ARE IN MM

ANGLE OF PROJECTION

100

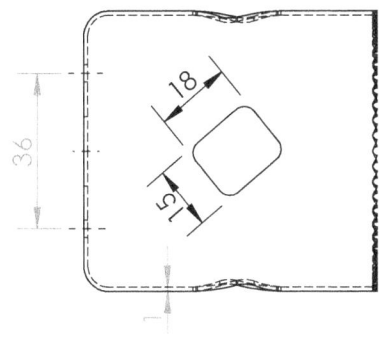
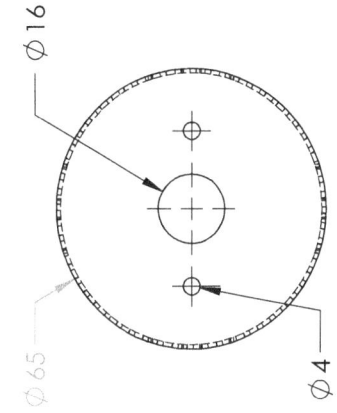
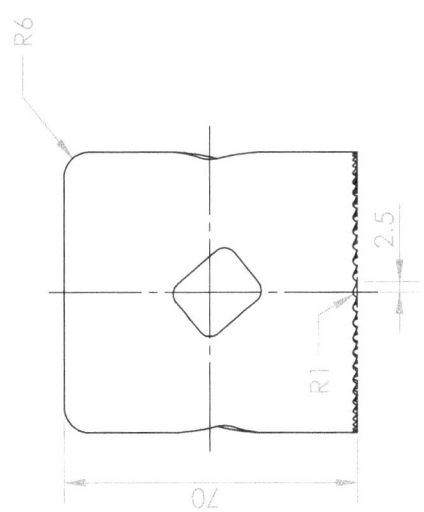
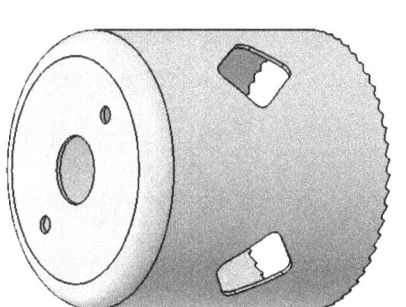

ALL DIMENSIONS ARE IN MM

ANGLE OF PROJECTION

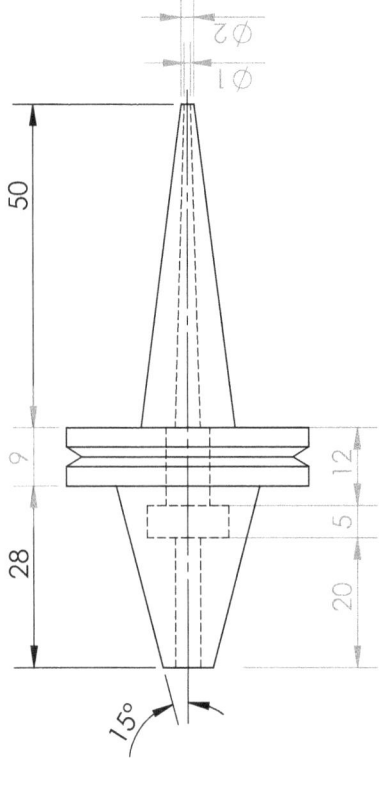
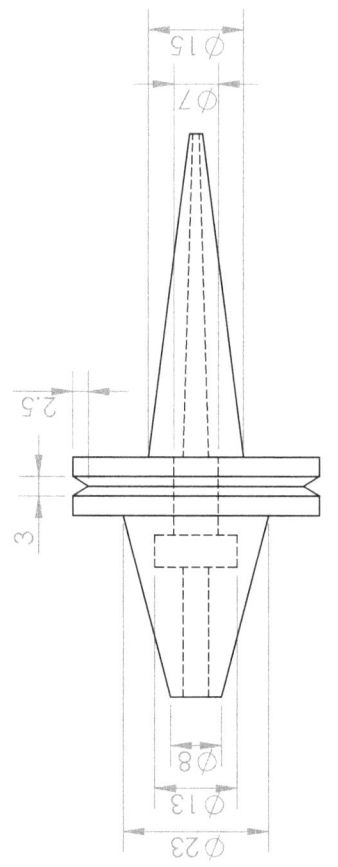
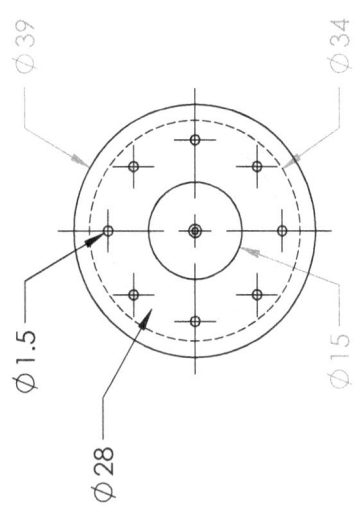

ALL DIMENSIONS ARE IN MM

ANGLE OF PROJECTION

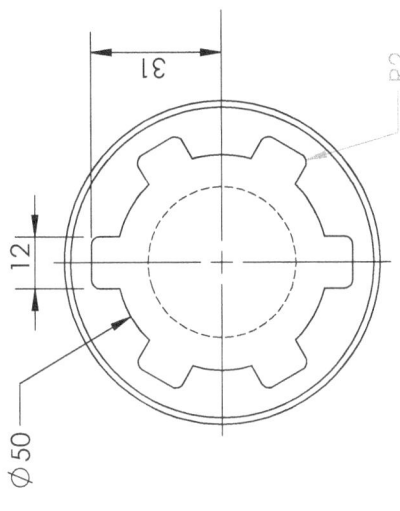
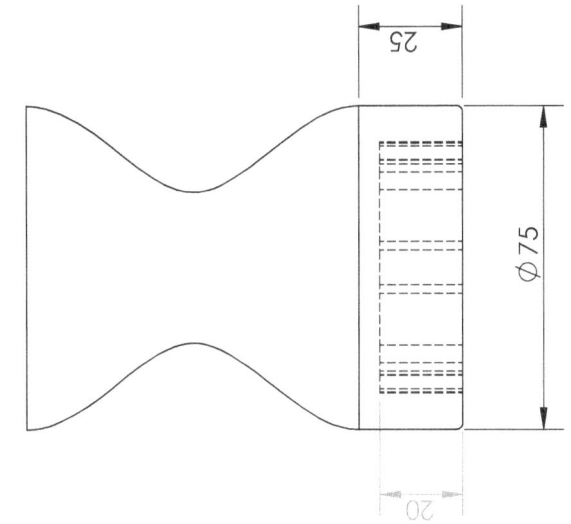
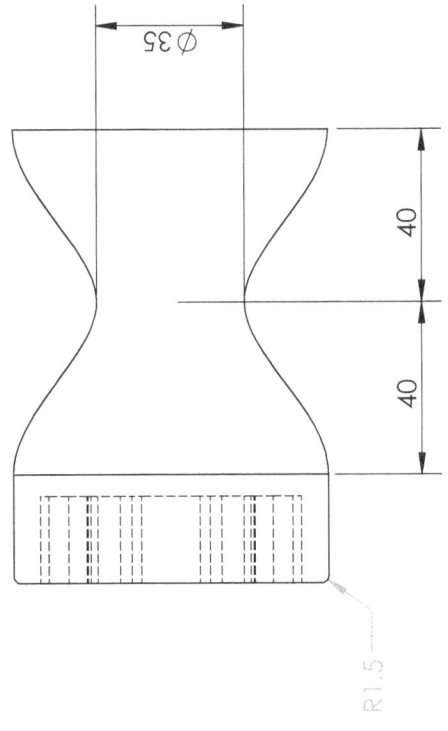
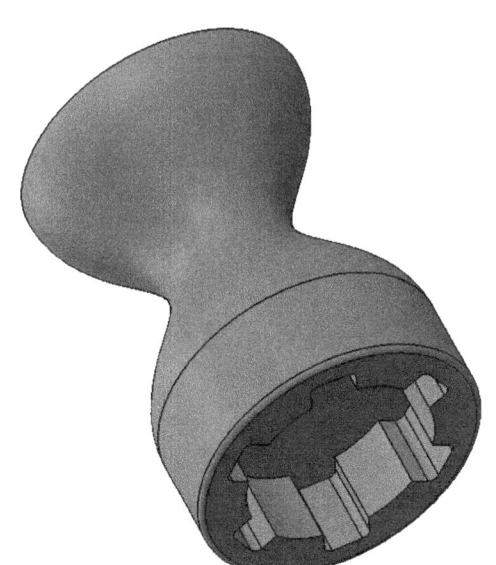

ALL DIMENSIONS ARE IN MM

ANGLE OF PROJECTION

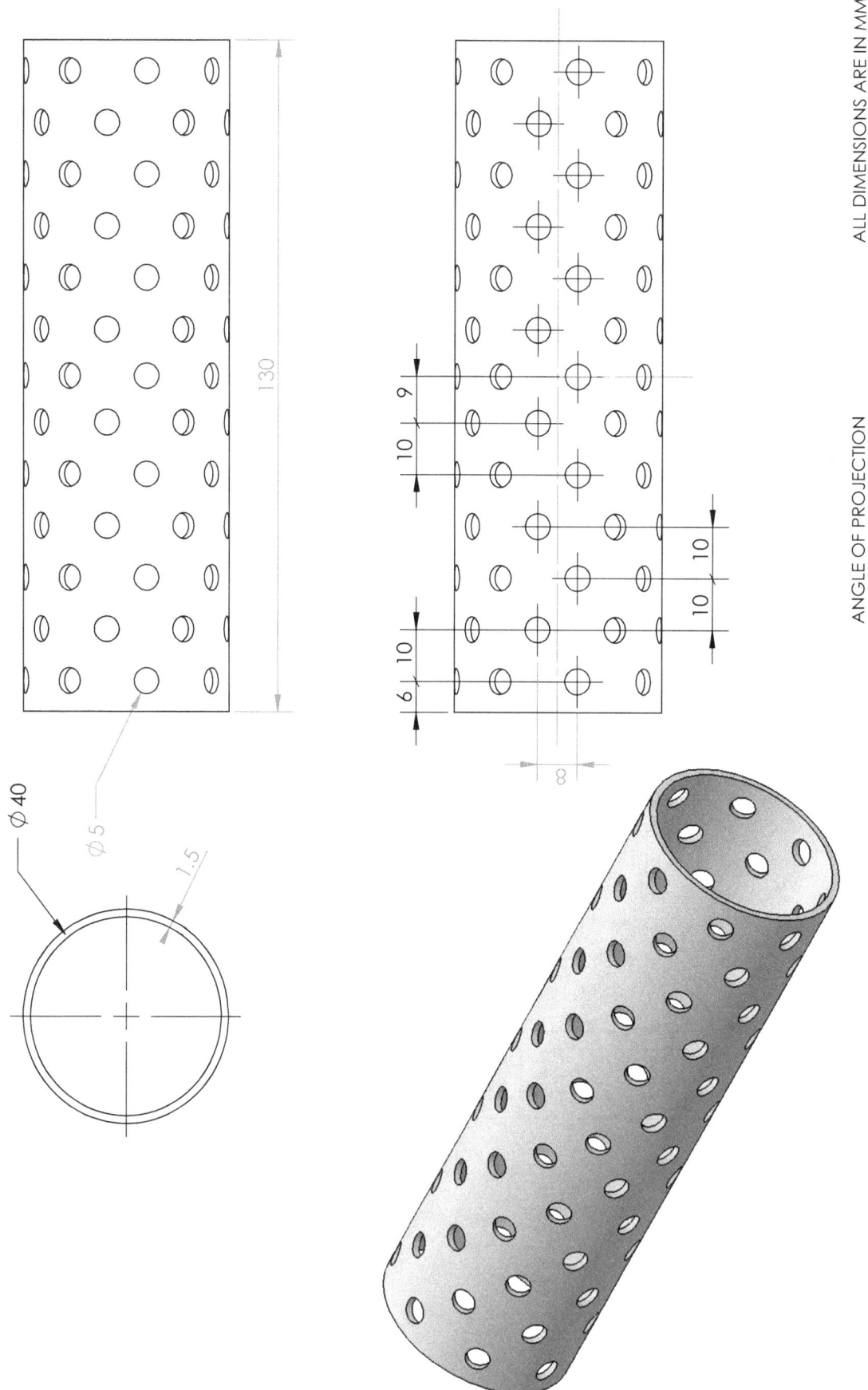

ALL DIMENSIONS ARE IN MM

ANGLE OF PROJECTION

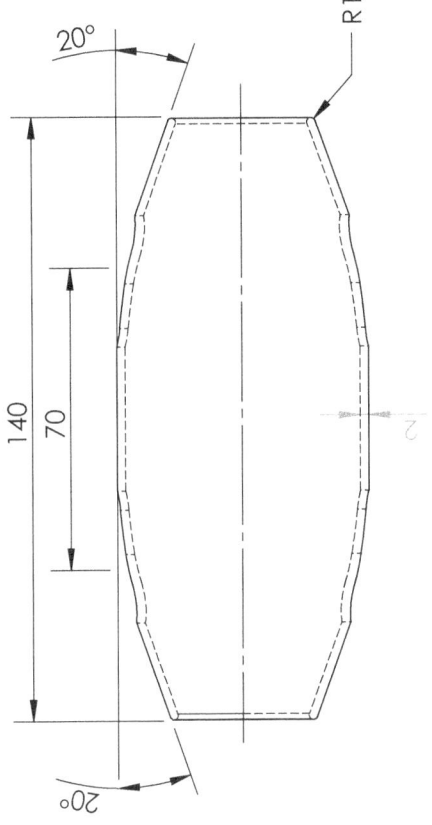
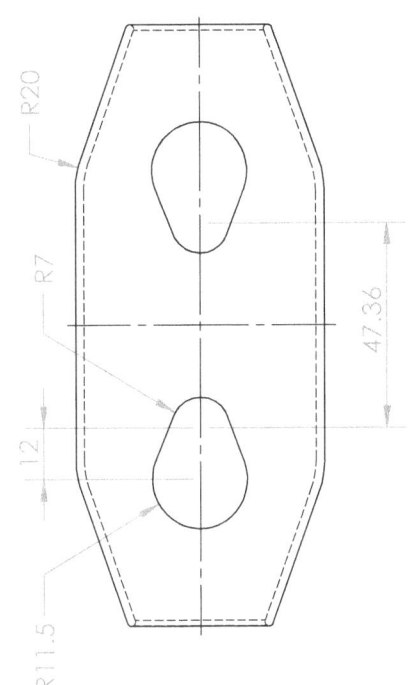
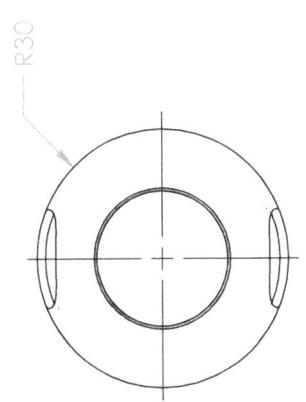
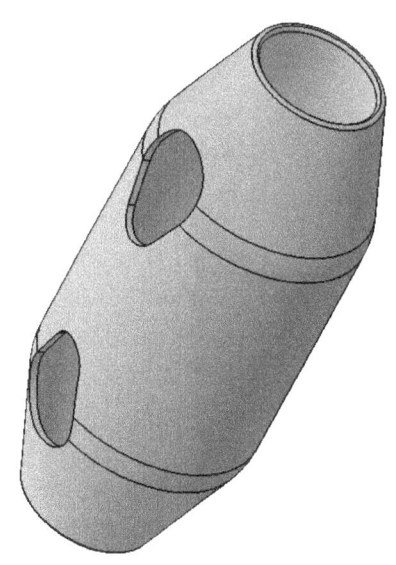

ALL DIMENSIONS ARE IN MM

ANGLE OF PROJECTION

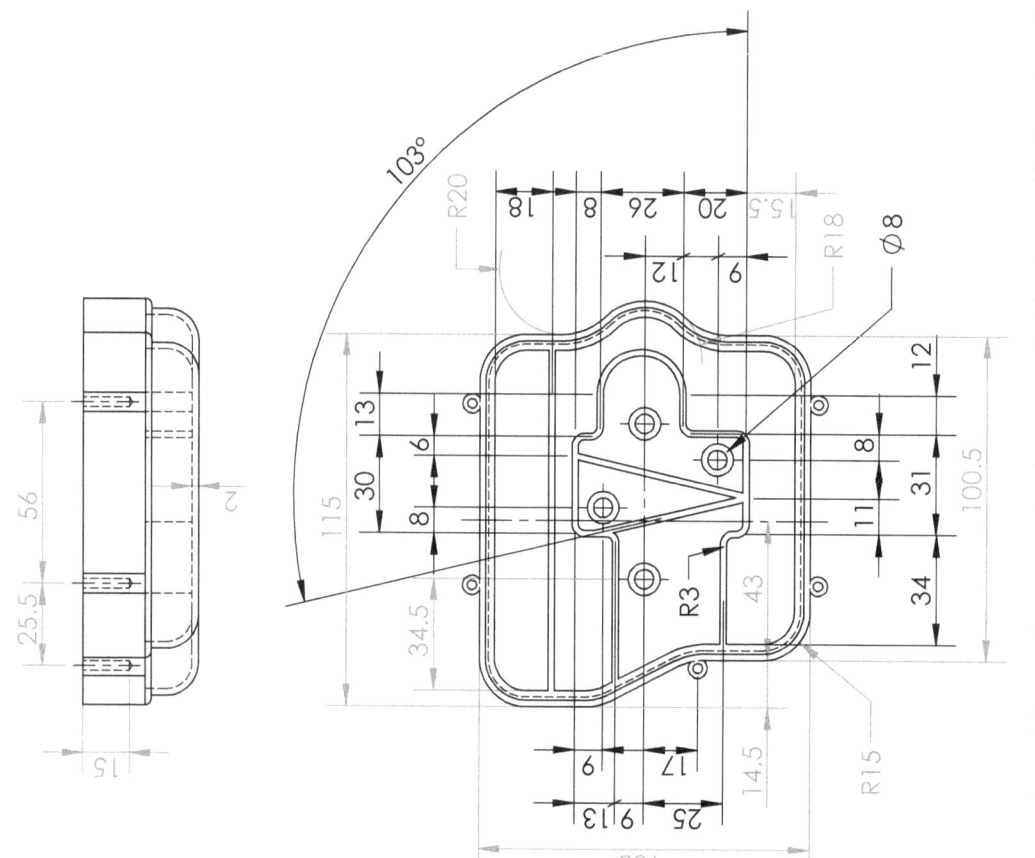
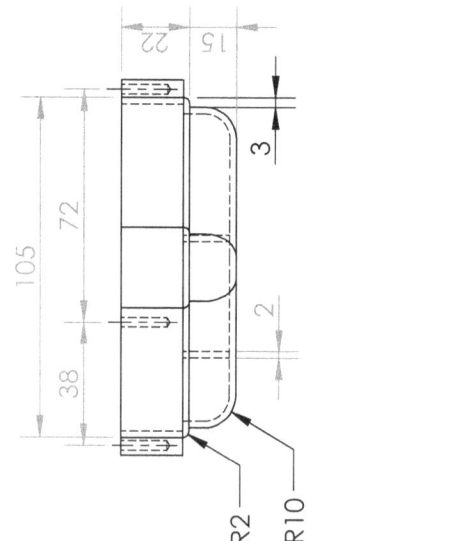
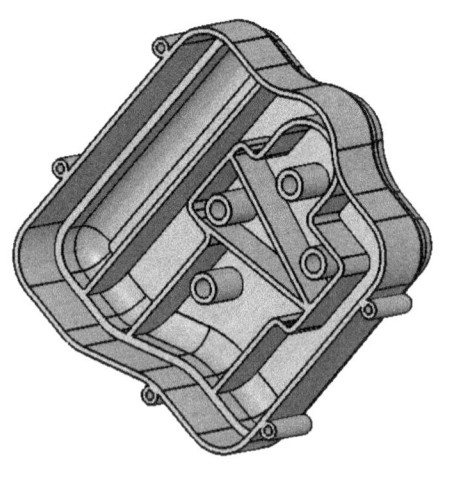

ALL DIMENSIONS ARE IN MM

ANGLE OF PROJECTION

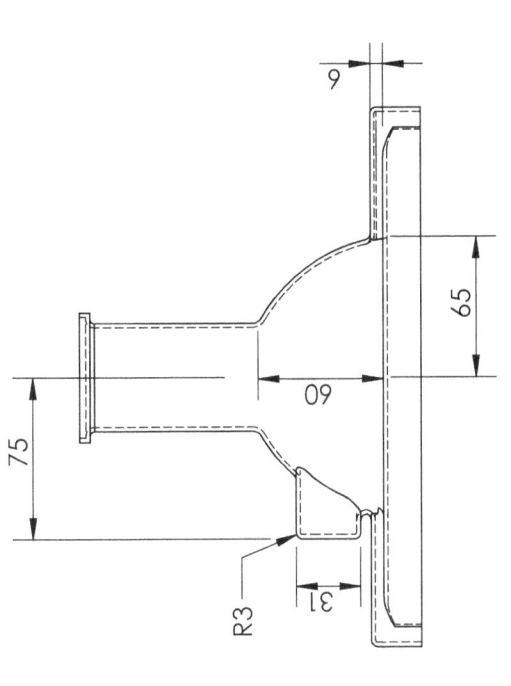
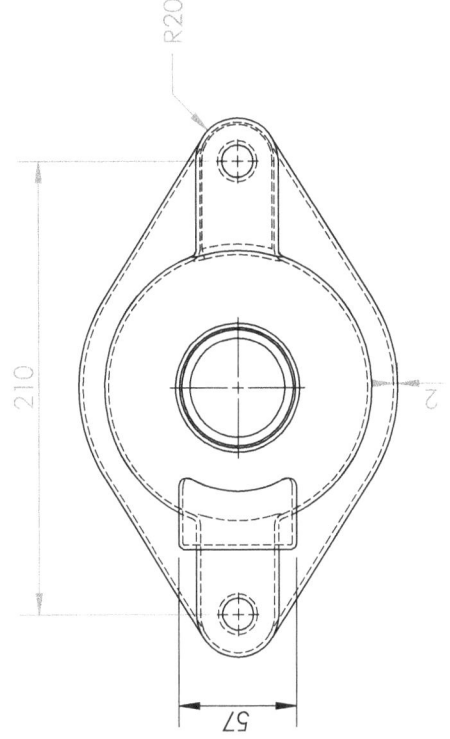
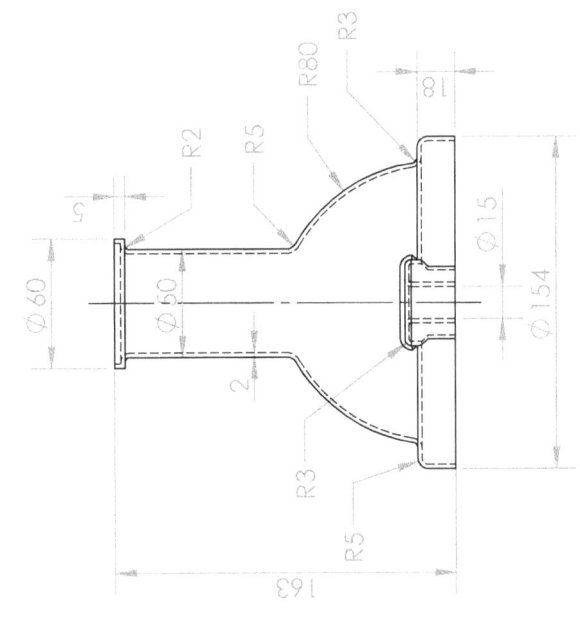
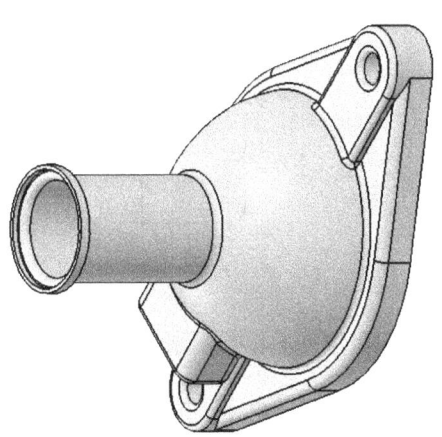

ALL DIMENSIONS ARE IN MM

ANGLE OF PROJECTION

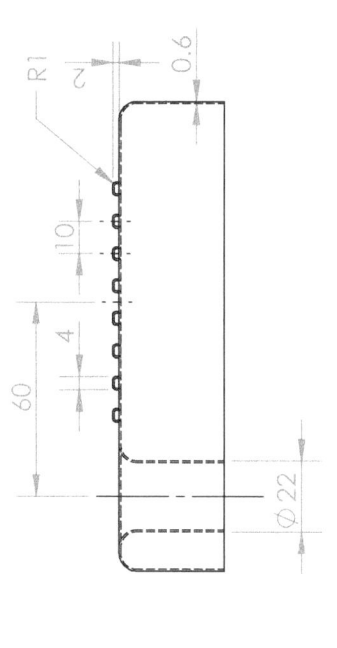
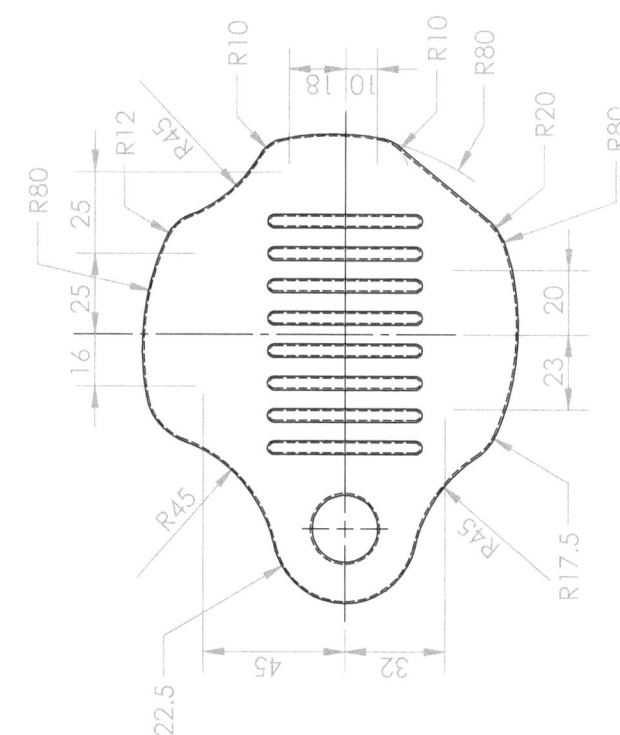
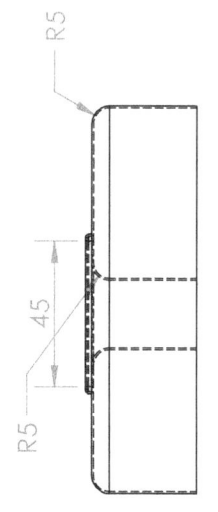
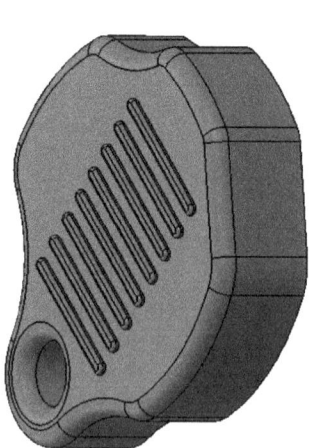

ALL DIMENSIONS ARE IN MM

ANGLE OF PROJECTION

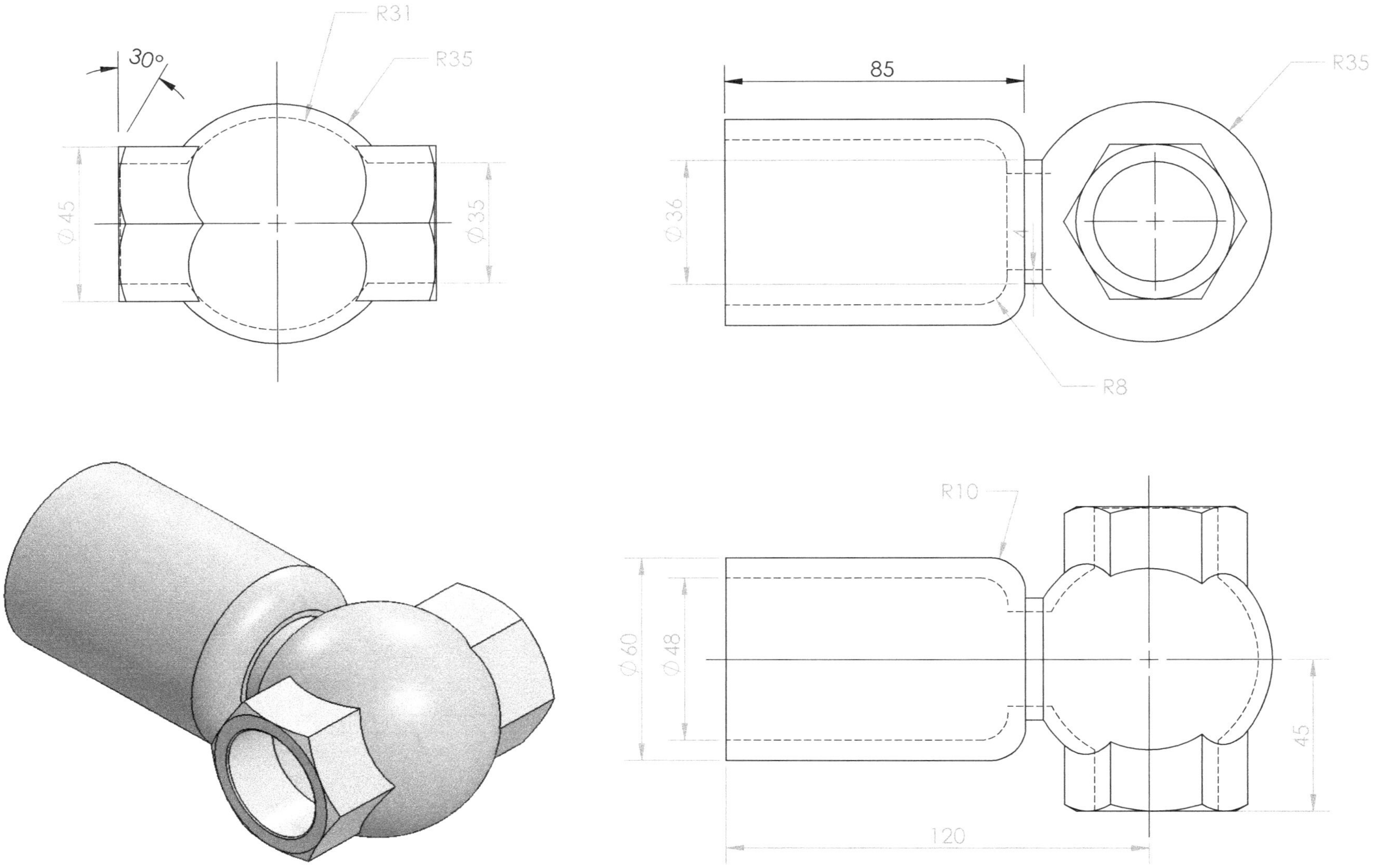

ALL DIMENSIONS ARE IN MM

ANGLE OF PROJECTION

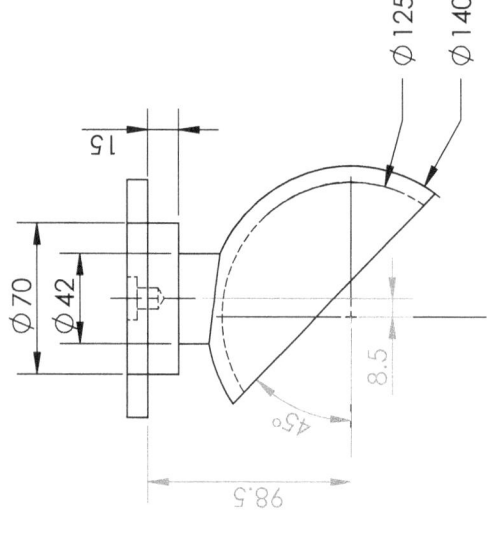
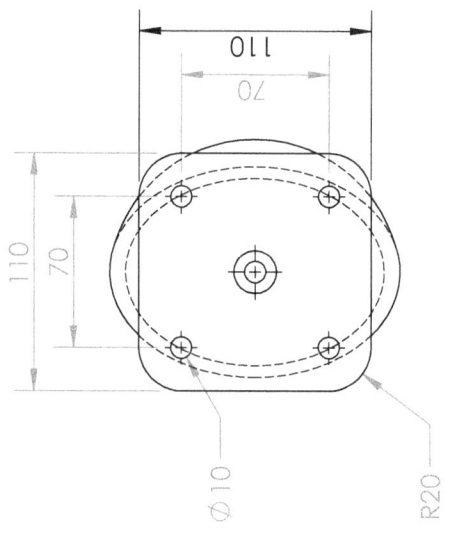
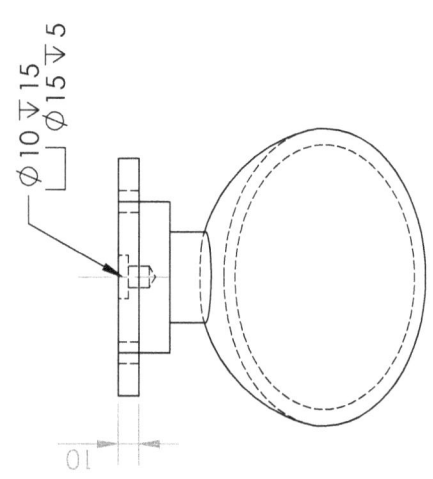
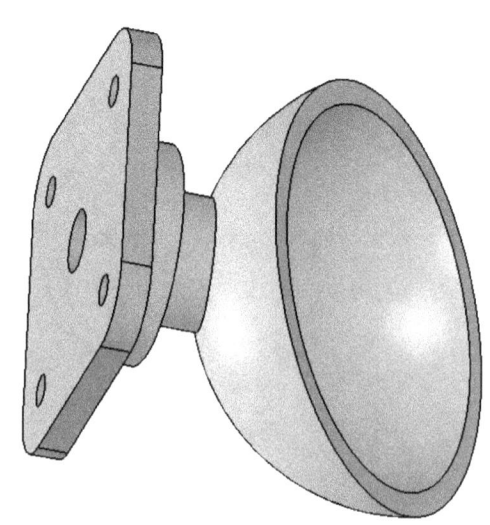

ALL DIMENSIONS ARE IN MM

ANGLE OF PROJECTION

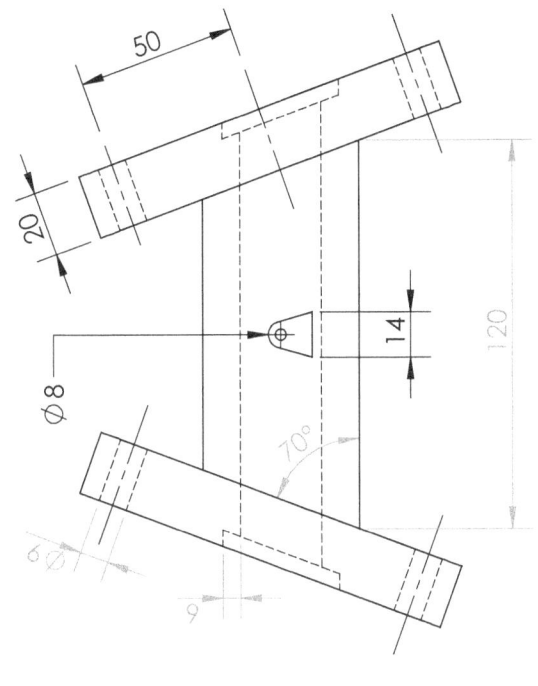
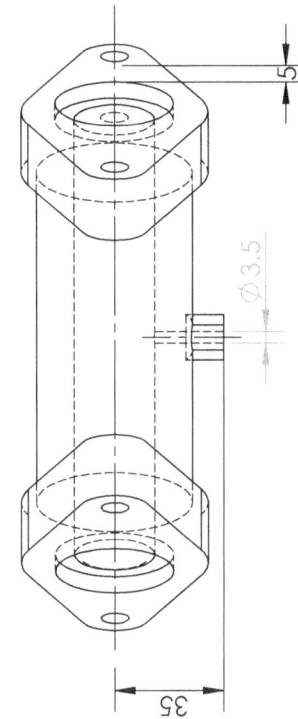
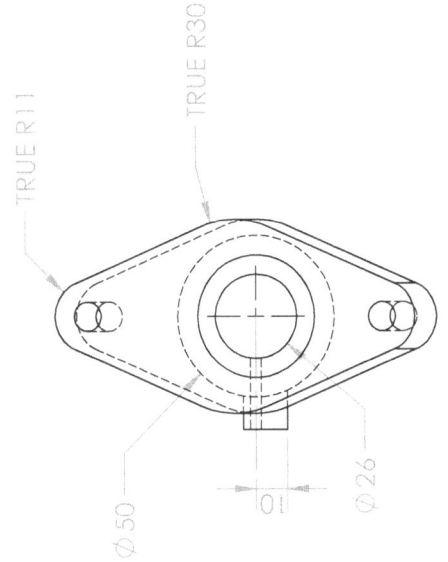
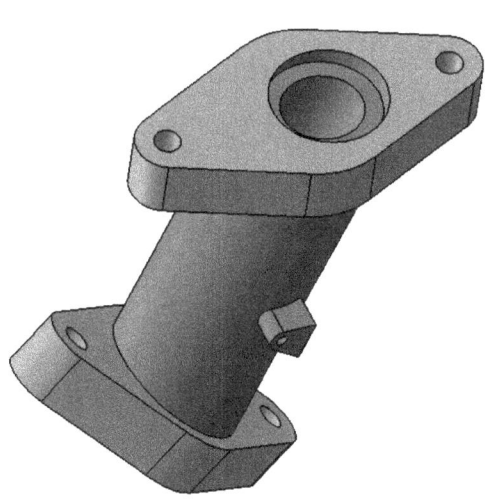

ALL DIMENSIONS ARE IN MM

ANGLE OF PROJECTION

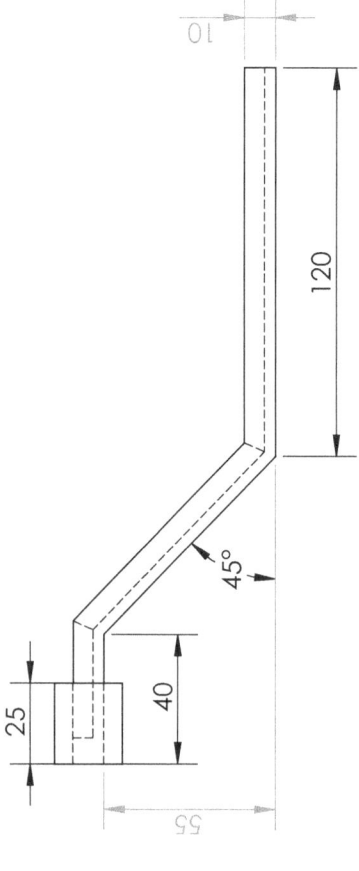
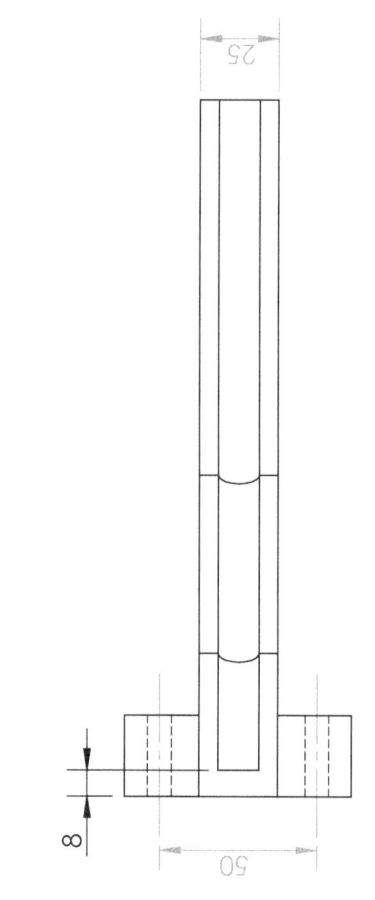
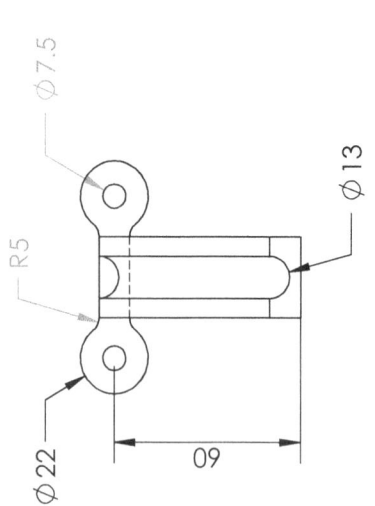

ALL DIMENSIONS ARE IN MM

ANGLE OF PROJECTION

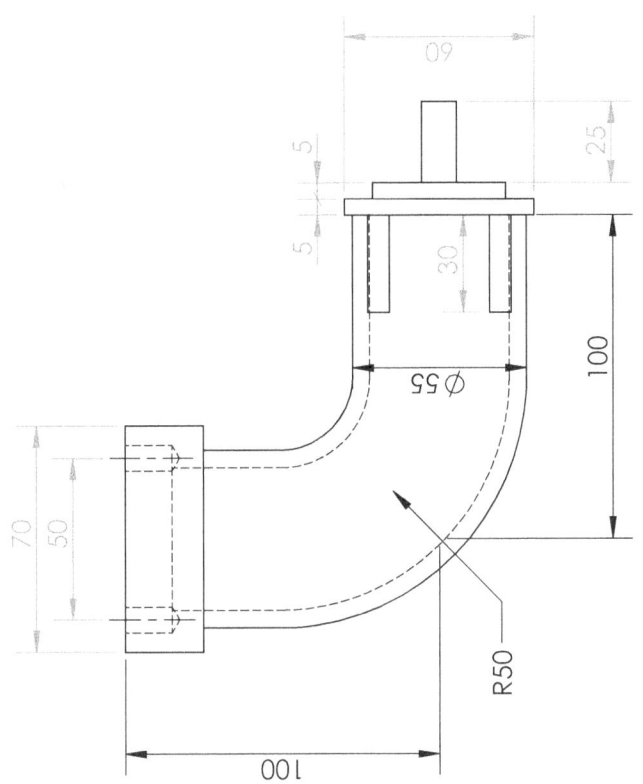
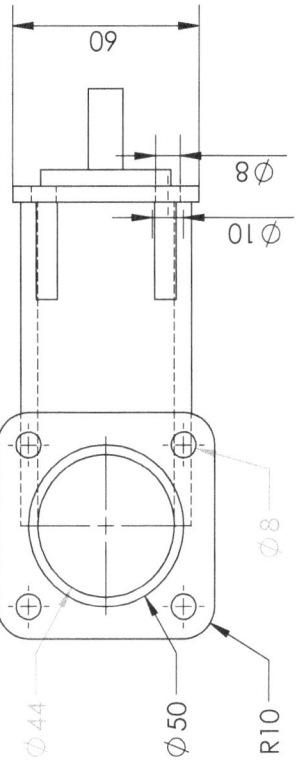
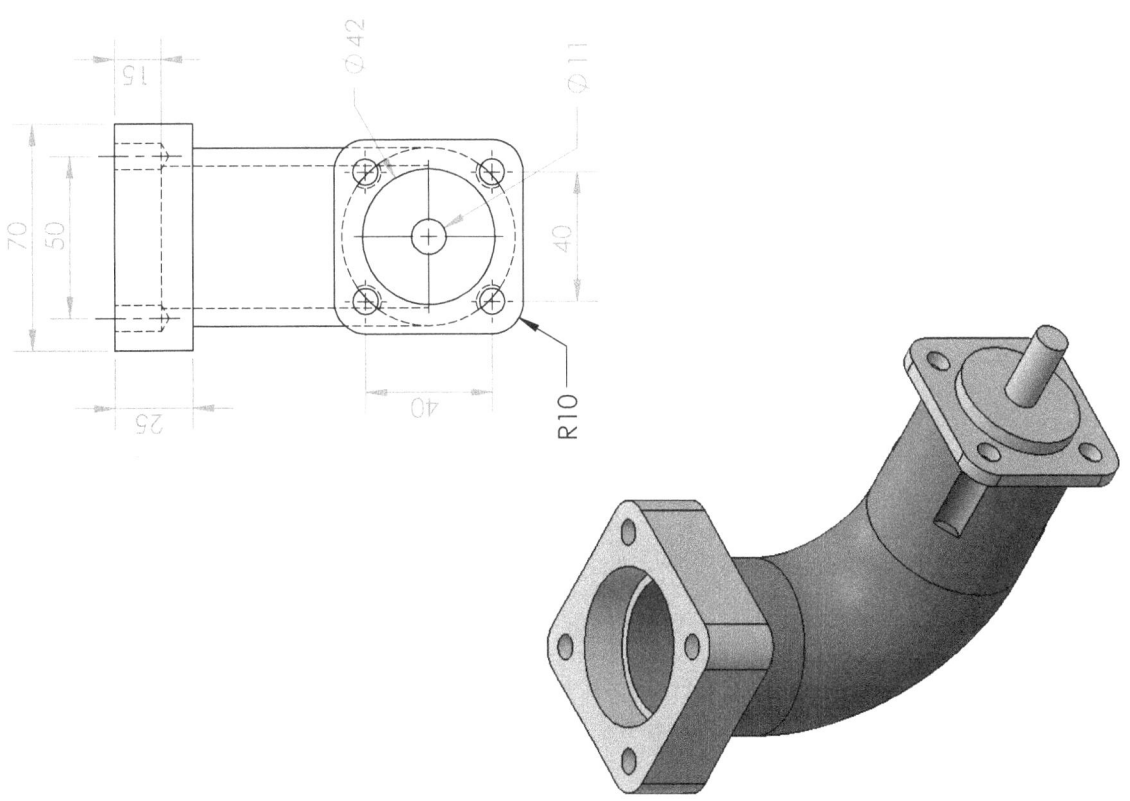

ALL DIMENSIONS ARE IN MM

ANGLE OF PROJECTION

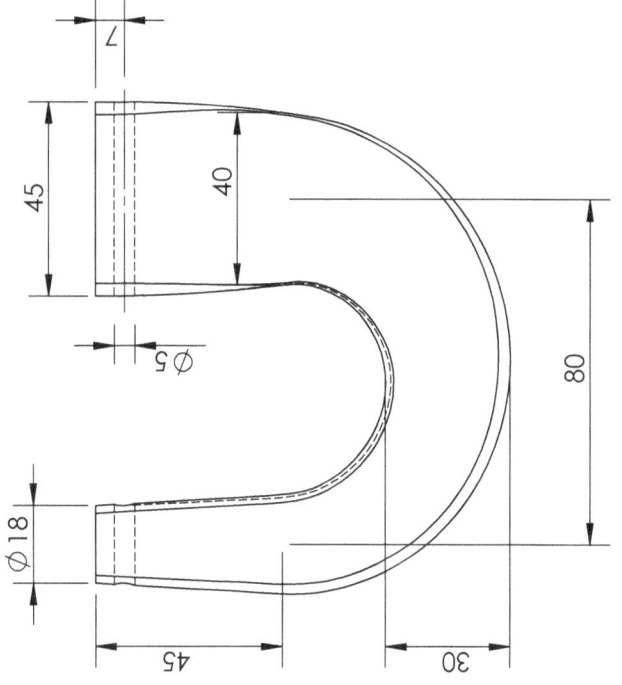
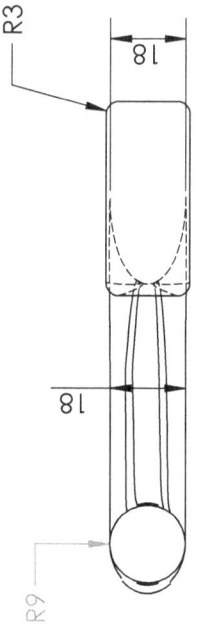
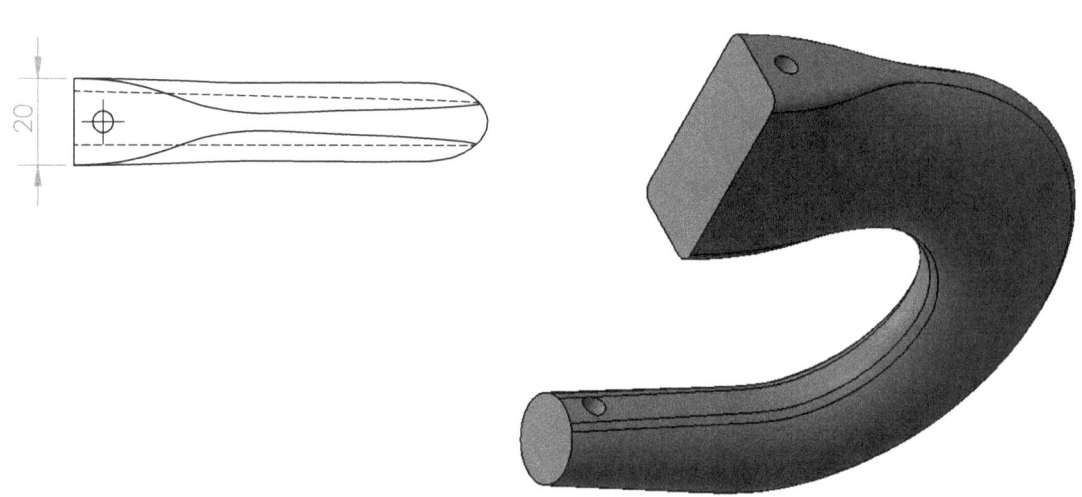

ALL DIMENSIONS ARE IN MM

ANGLE OF PROJECTION

ALL DIMENSIONS ARE IN MM

ANGLE OF PROJECTION

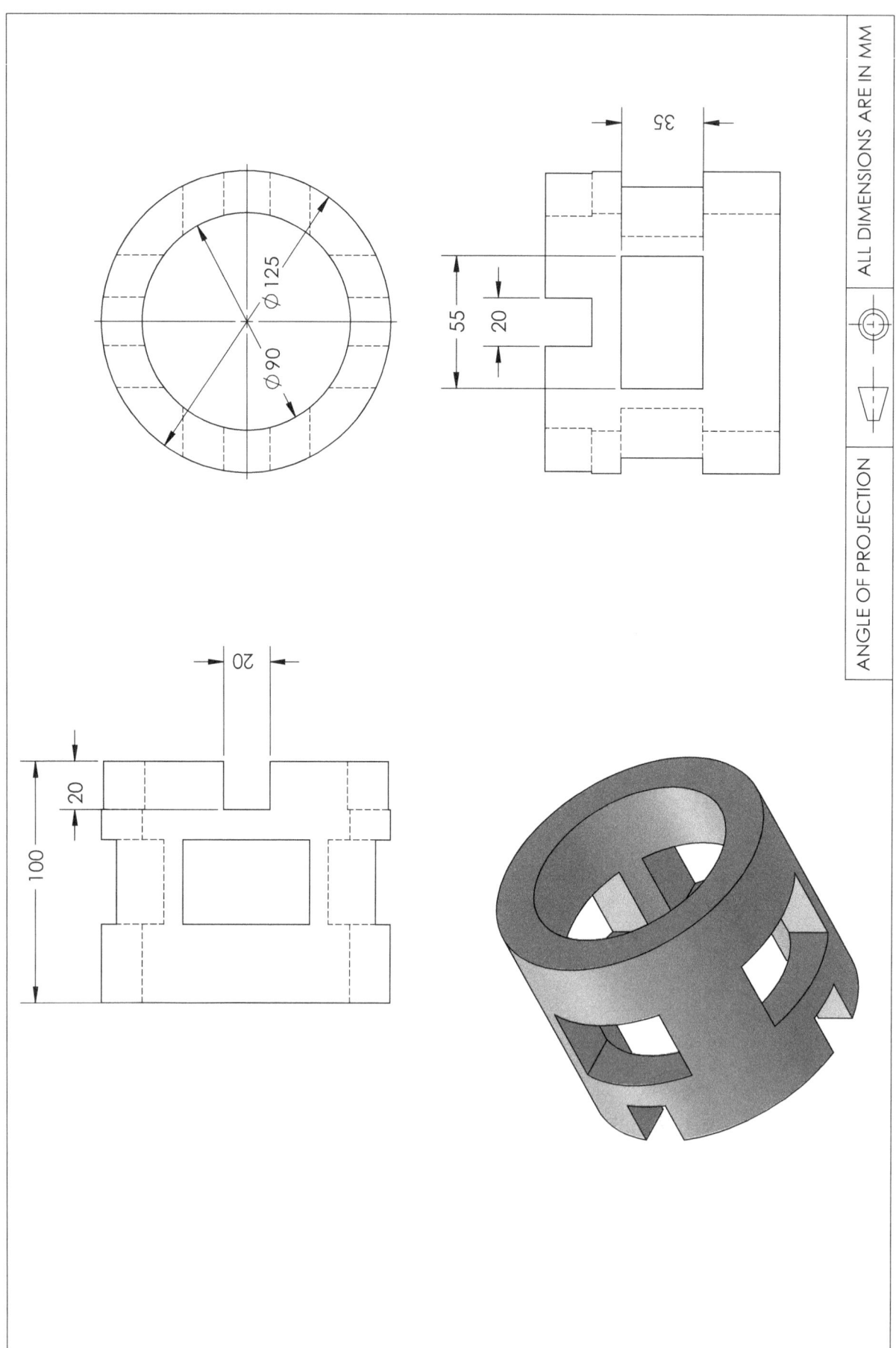

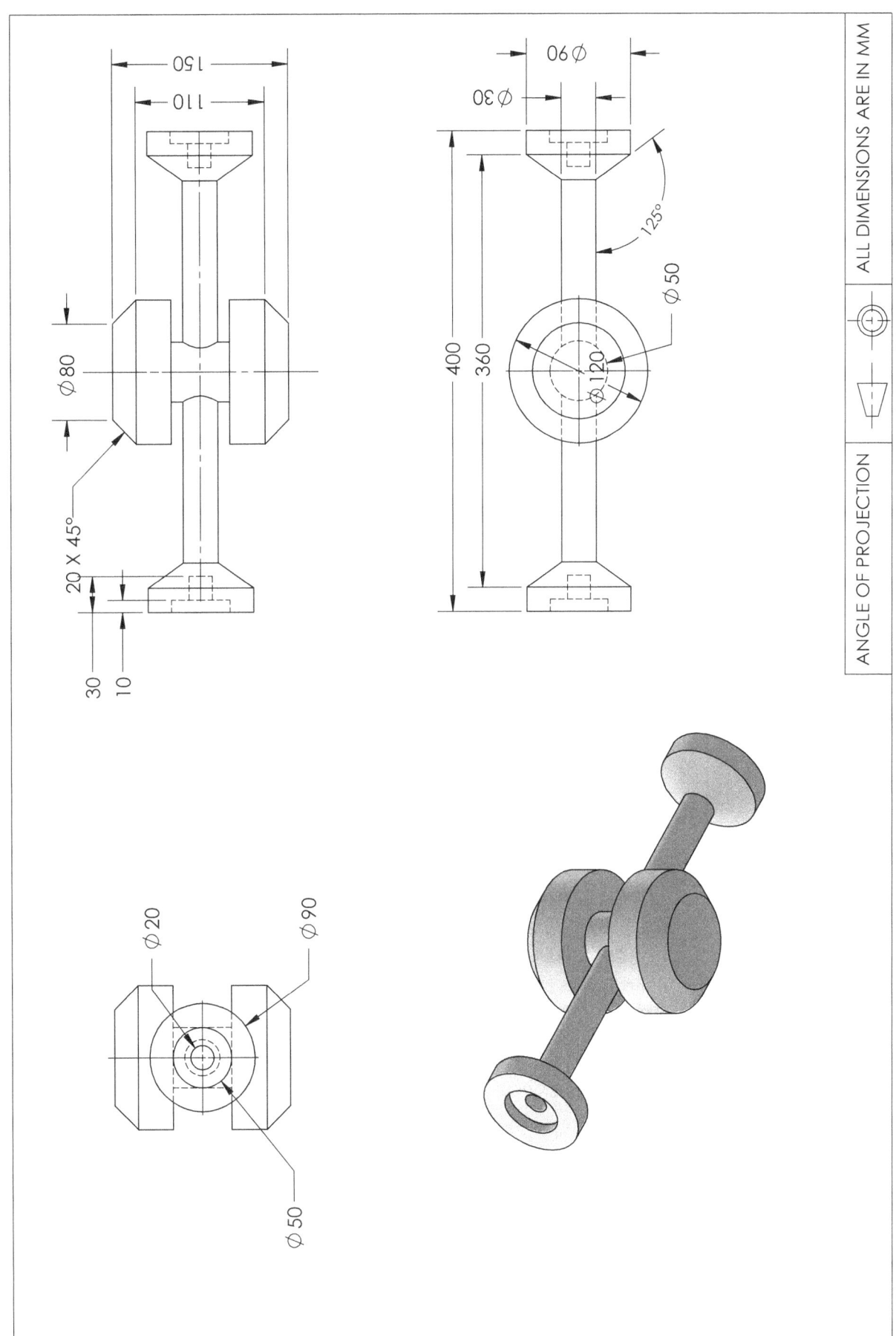

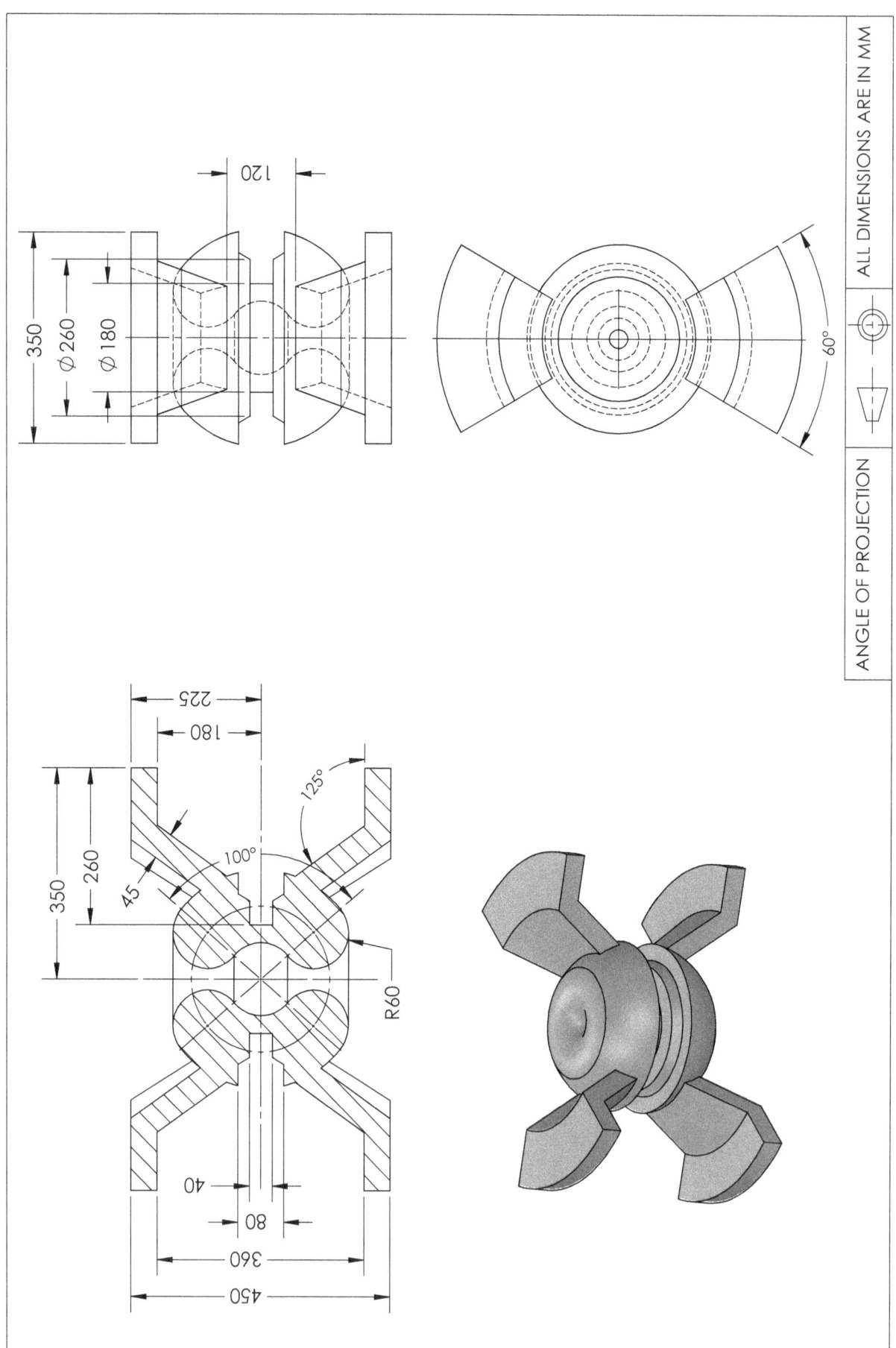

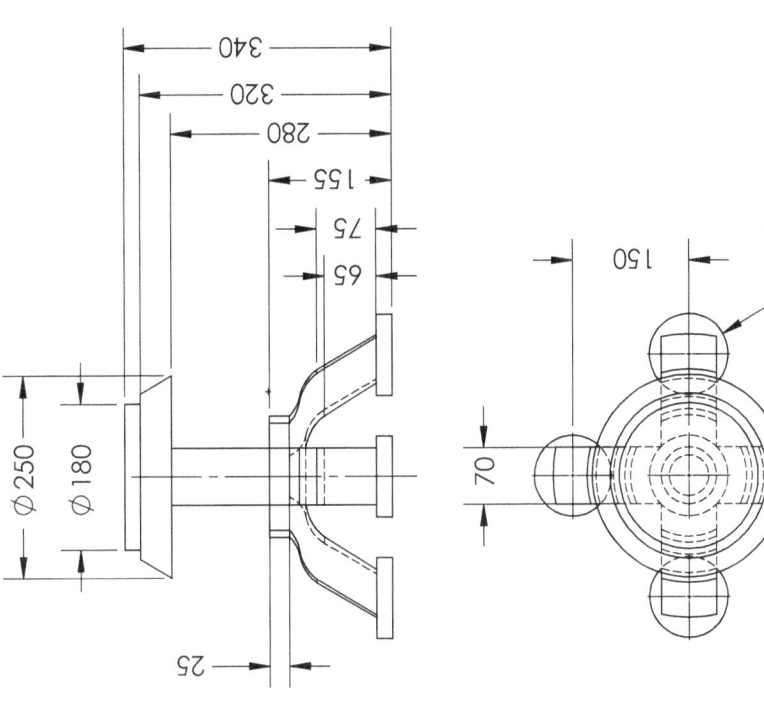
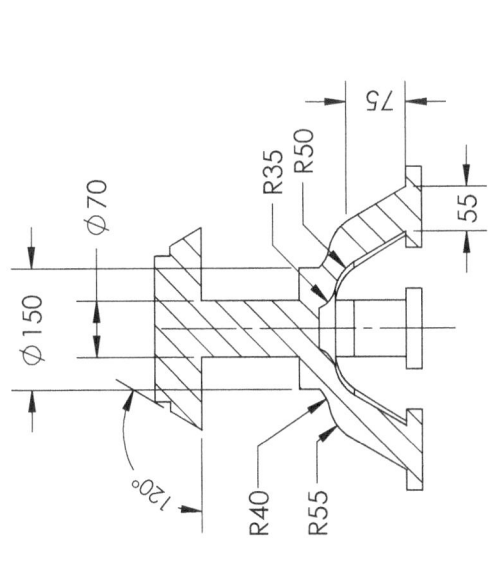
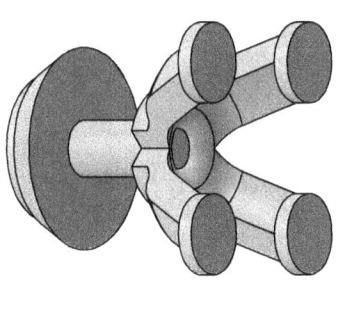

ALL DIMENSIONS ARE IN MM

ANGLE OF PROJECTION

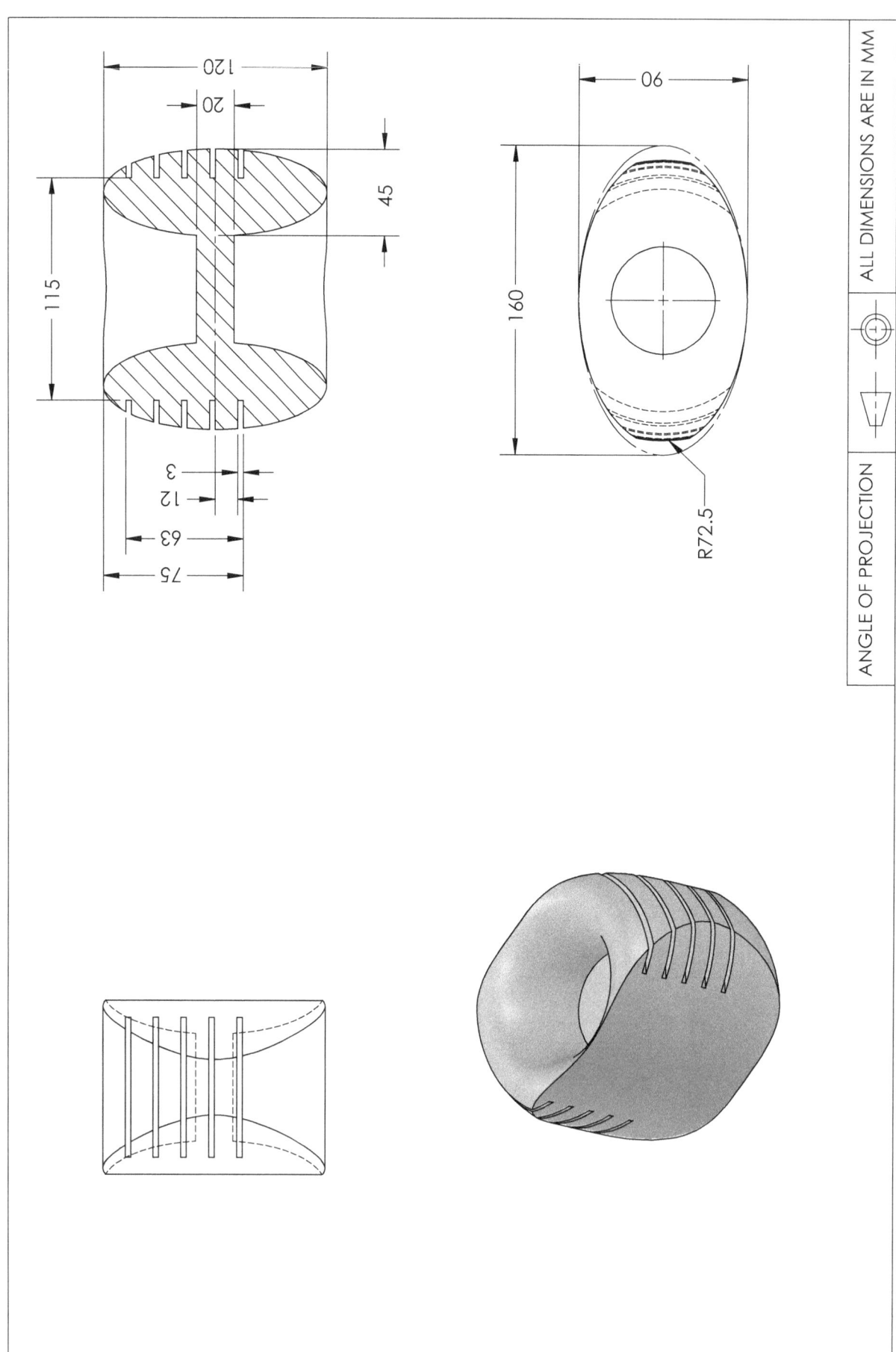

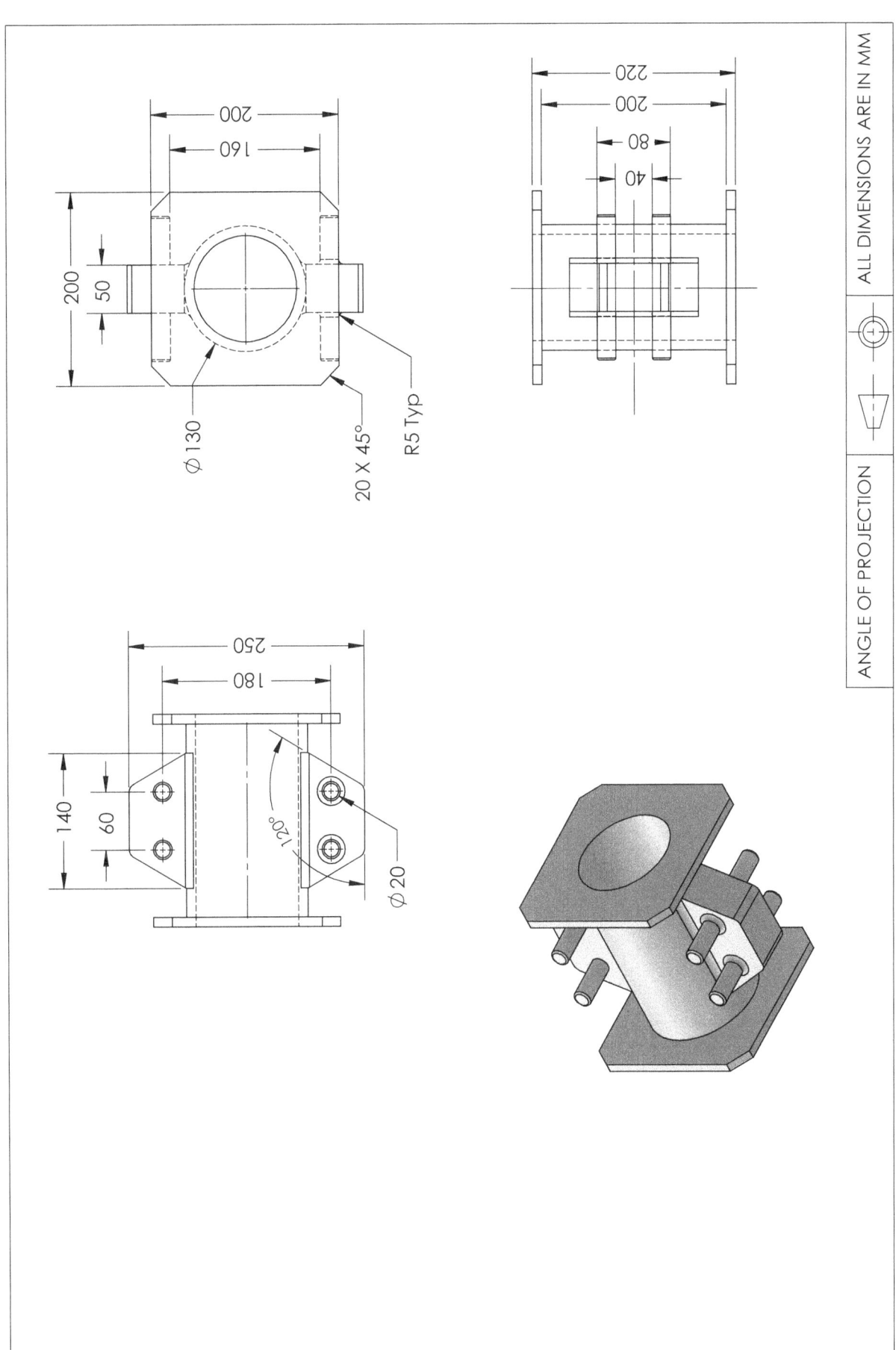

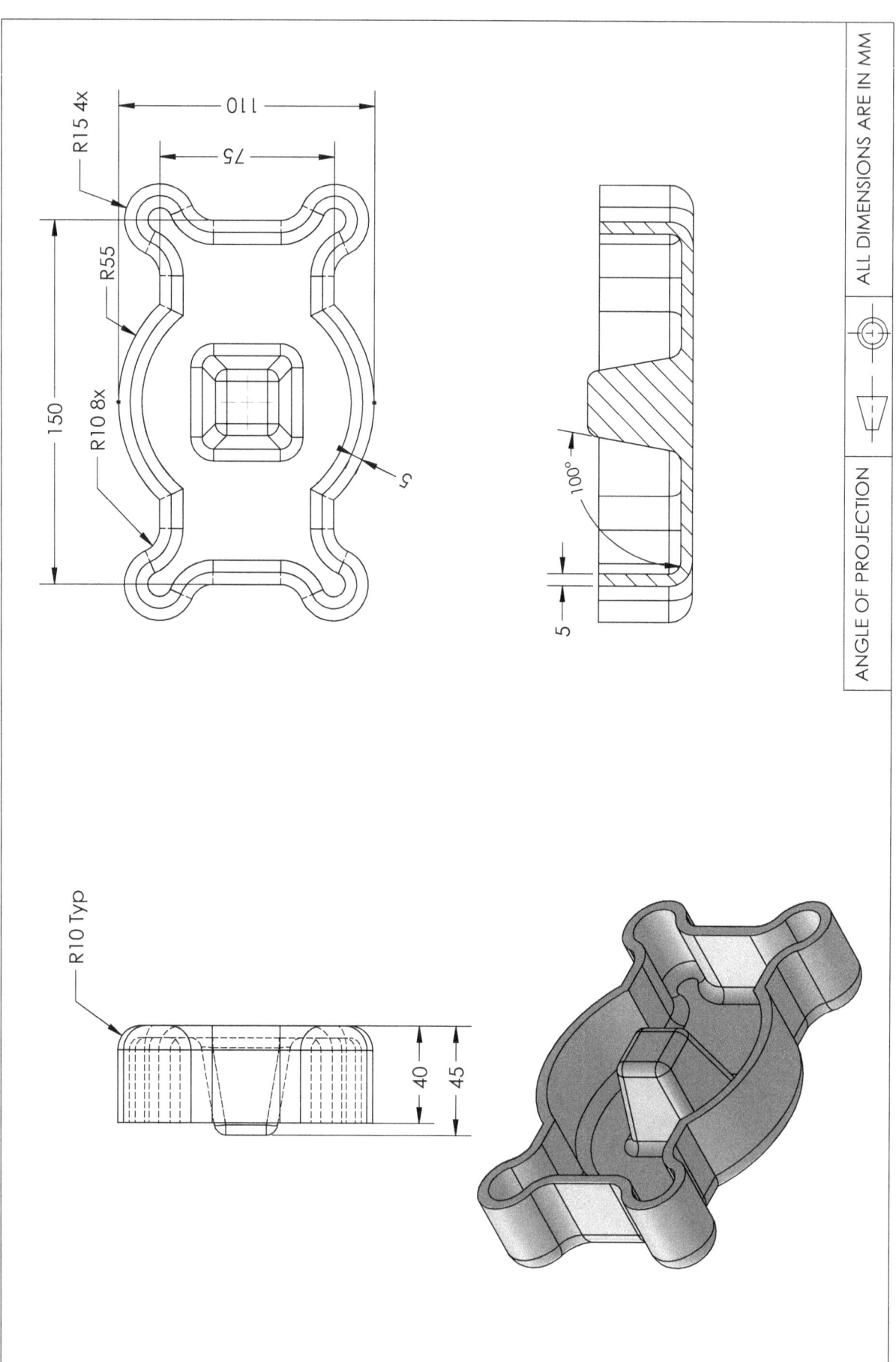

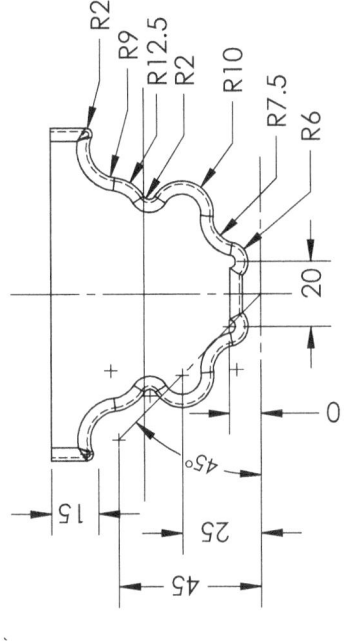
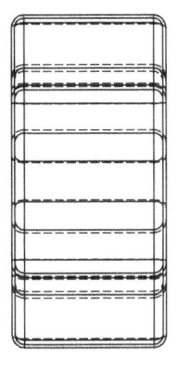
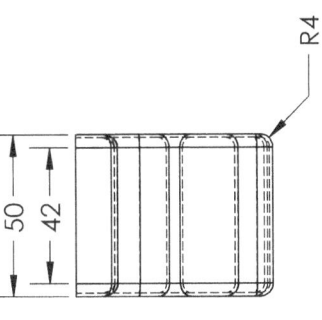
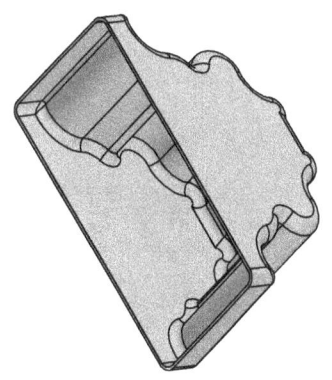

ANGLE OF PROJECTION   ALL DIMENSIONS ARE IN MM

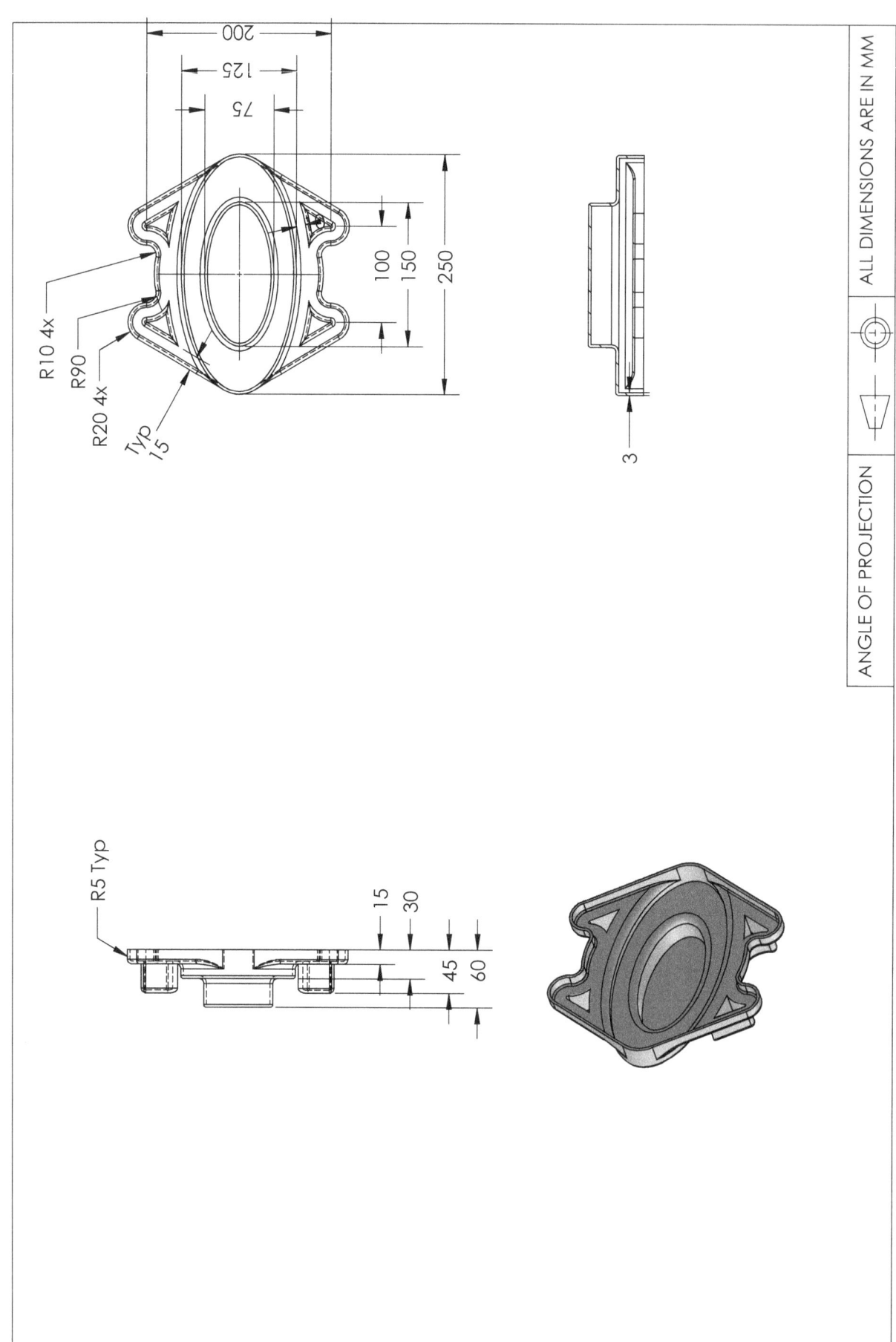

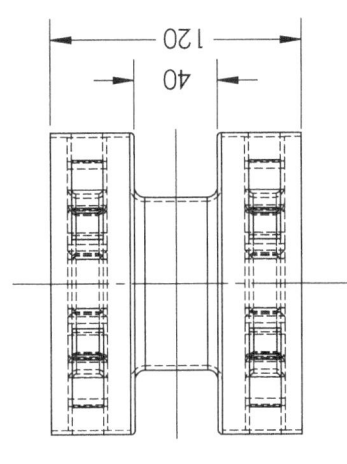
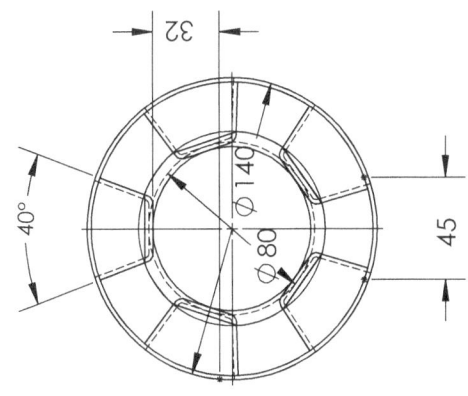
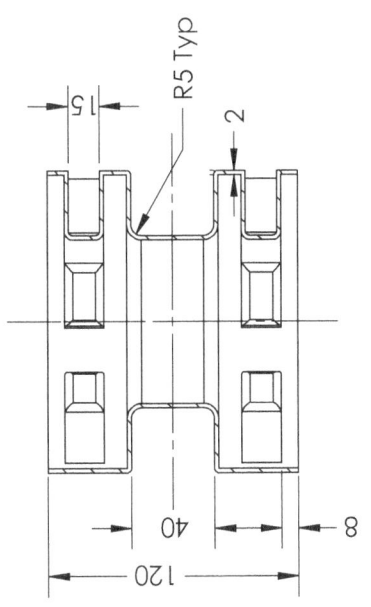

ALL DIMENSIONS ARE IN MM

ANGLE OF PROJECTION

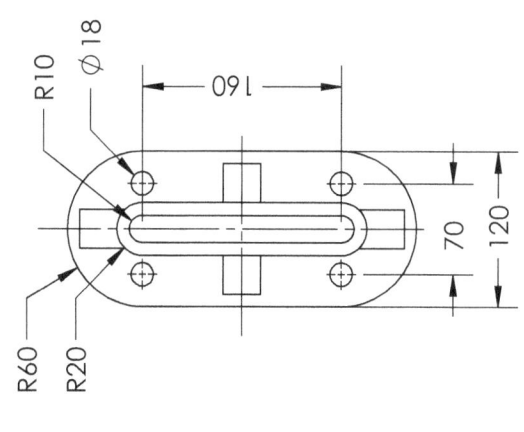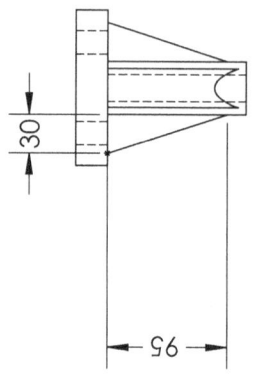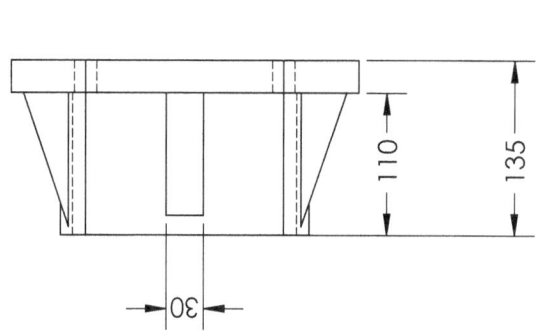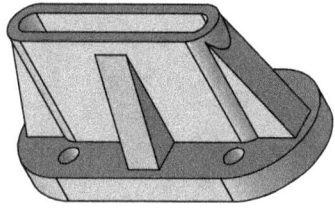

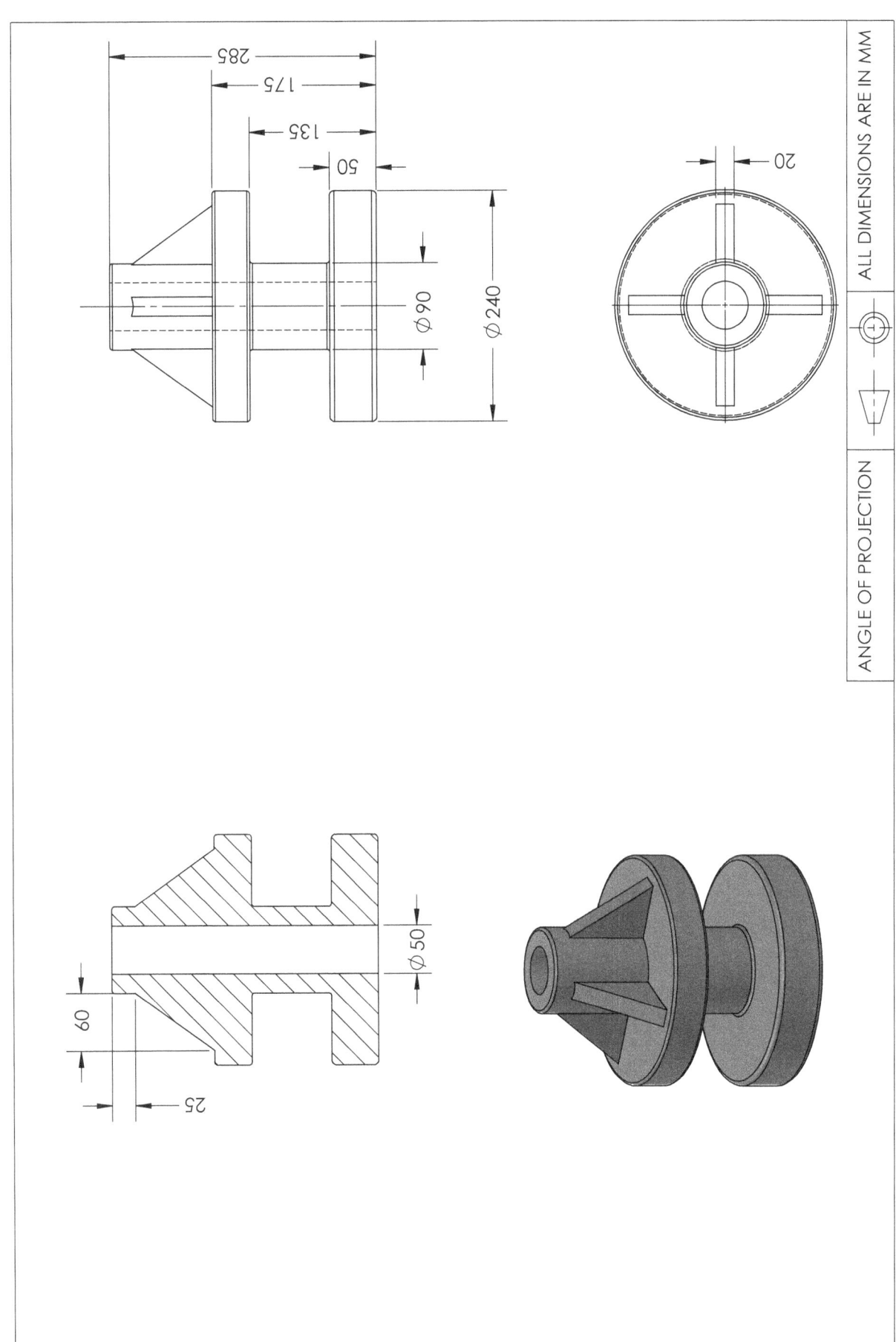

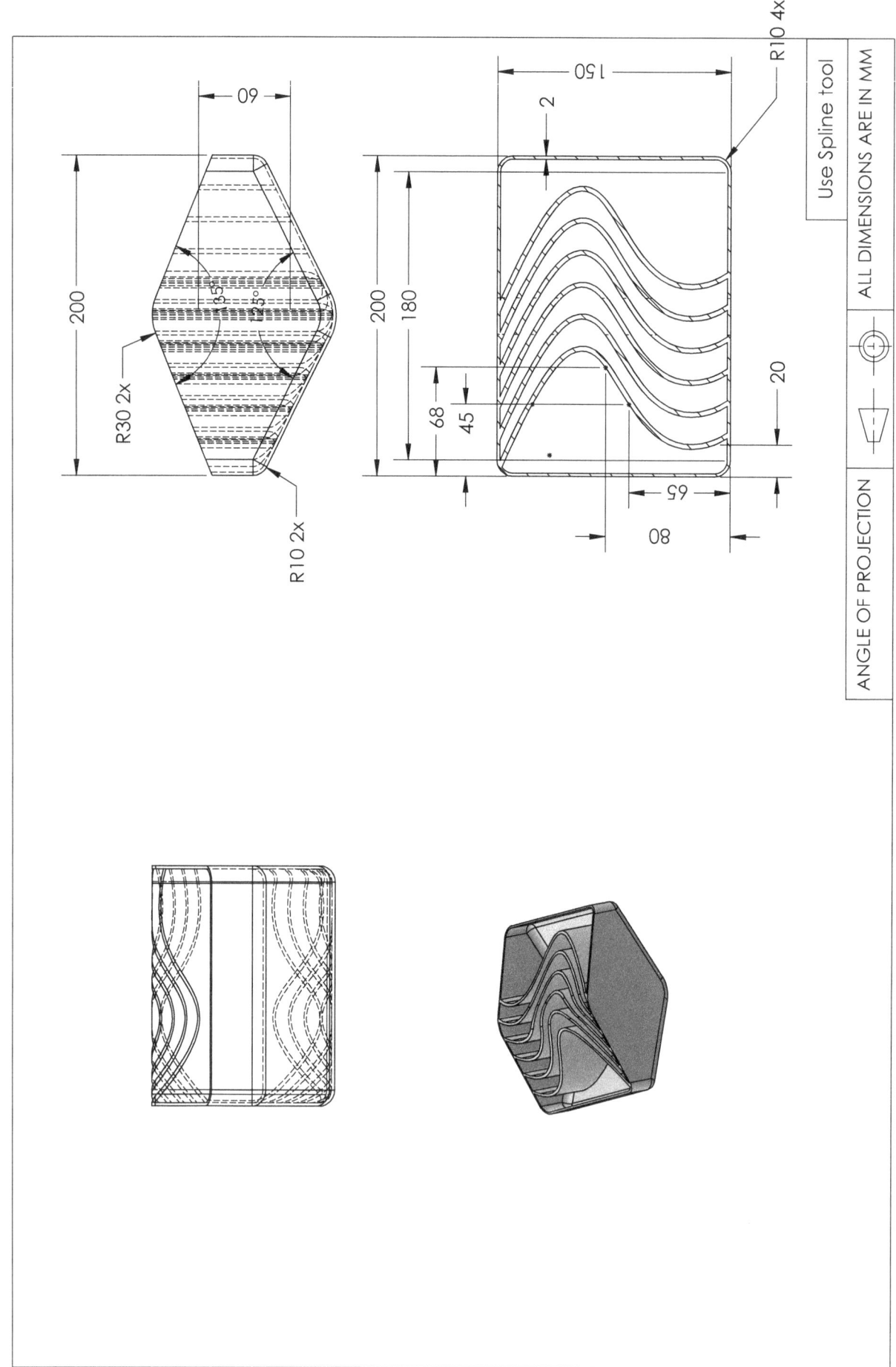

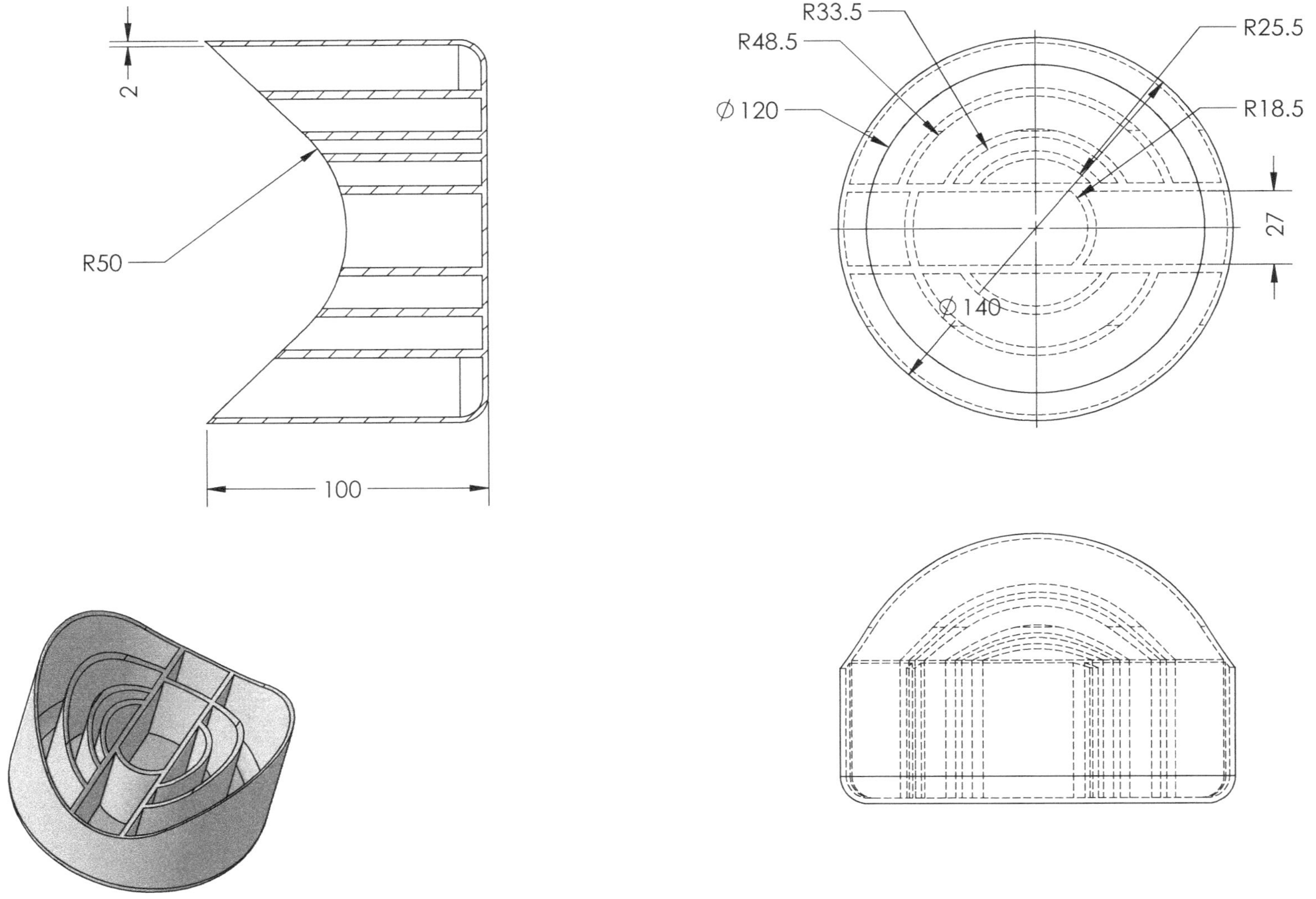

ALL DIMENSIONS ARE IN MM

ANGLE OF PROJECTION

R25.5
R18.5
27
Ø140
R33.5
R48.5
Ø120

100
2
R50

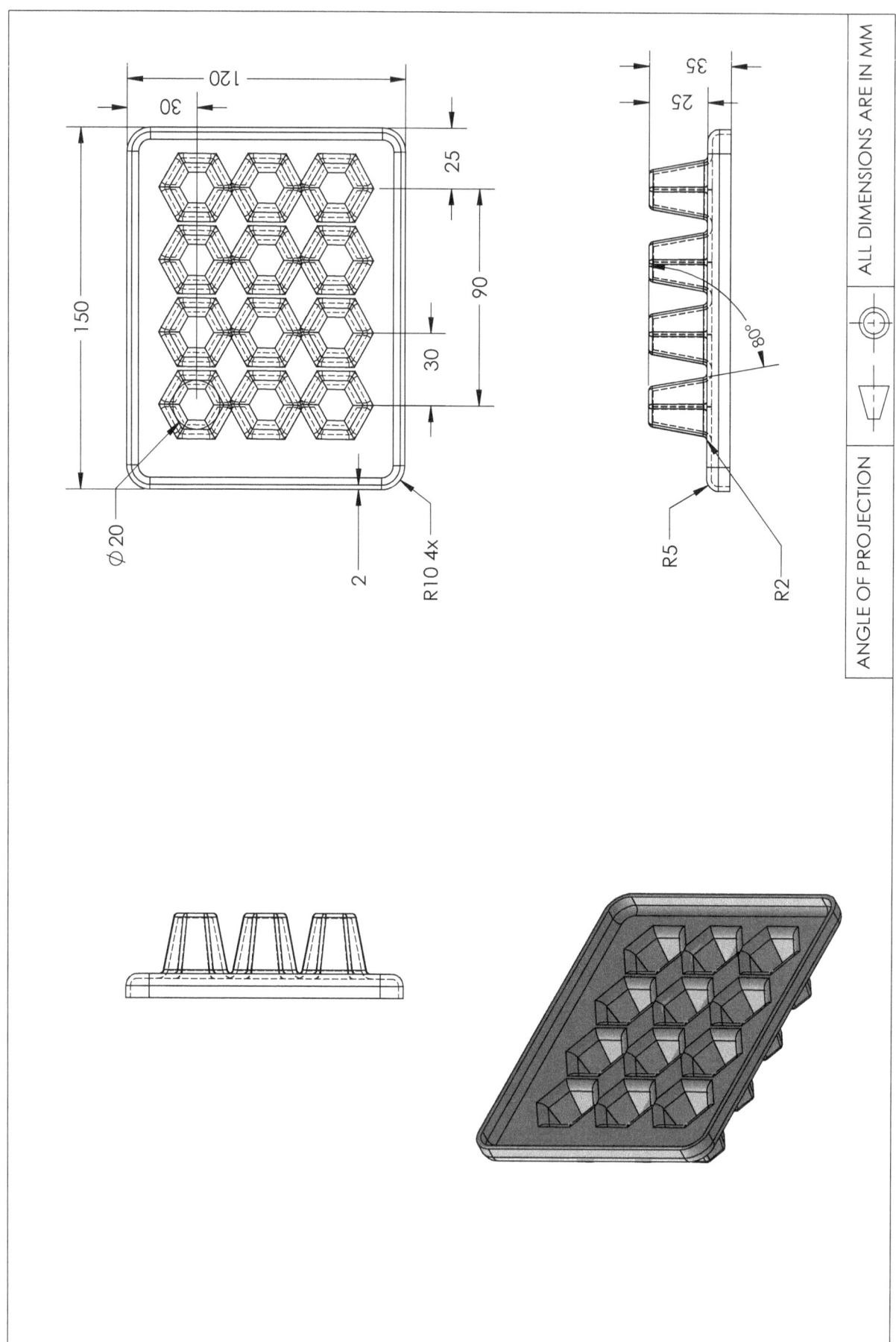

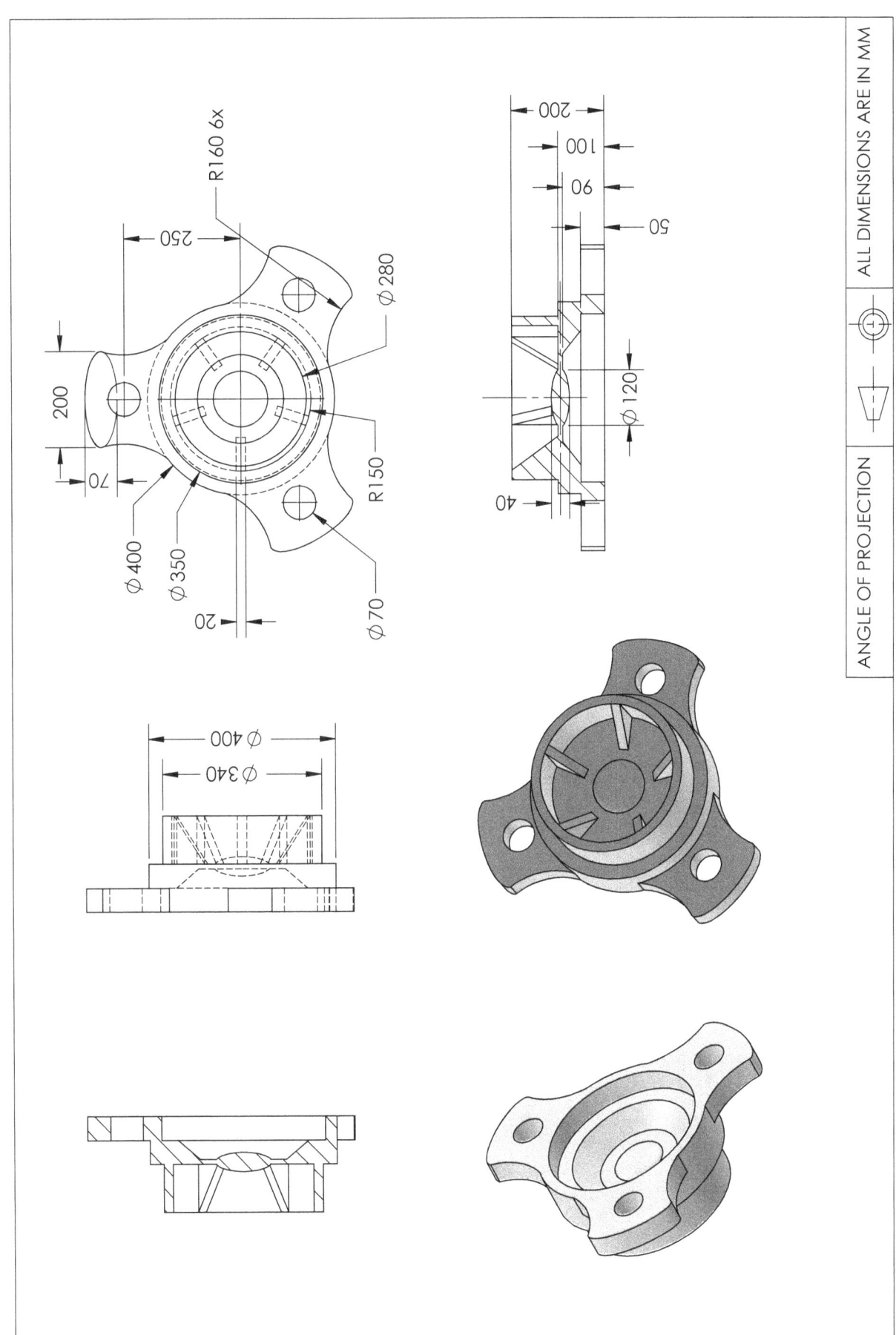

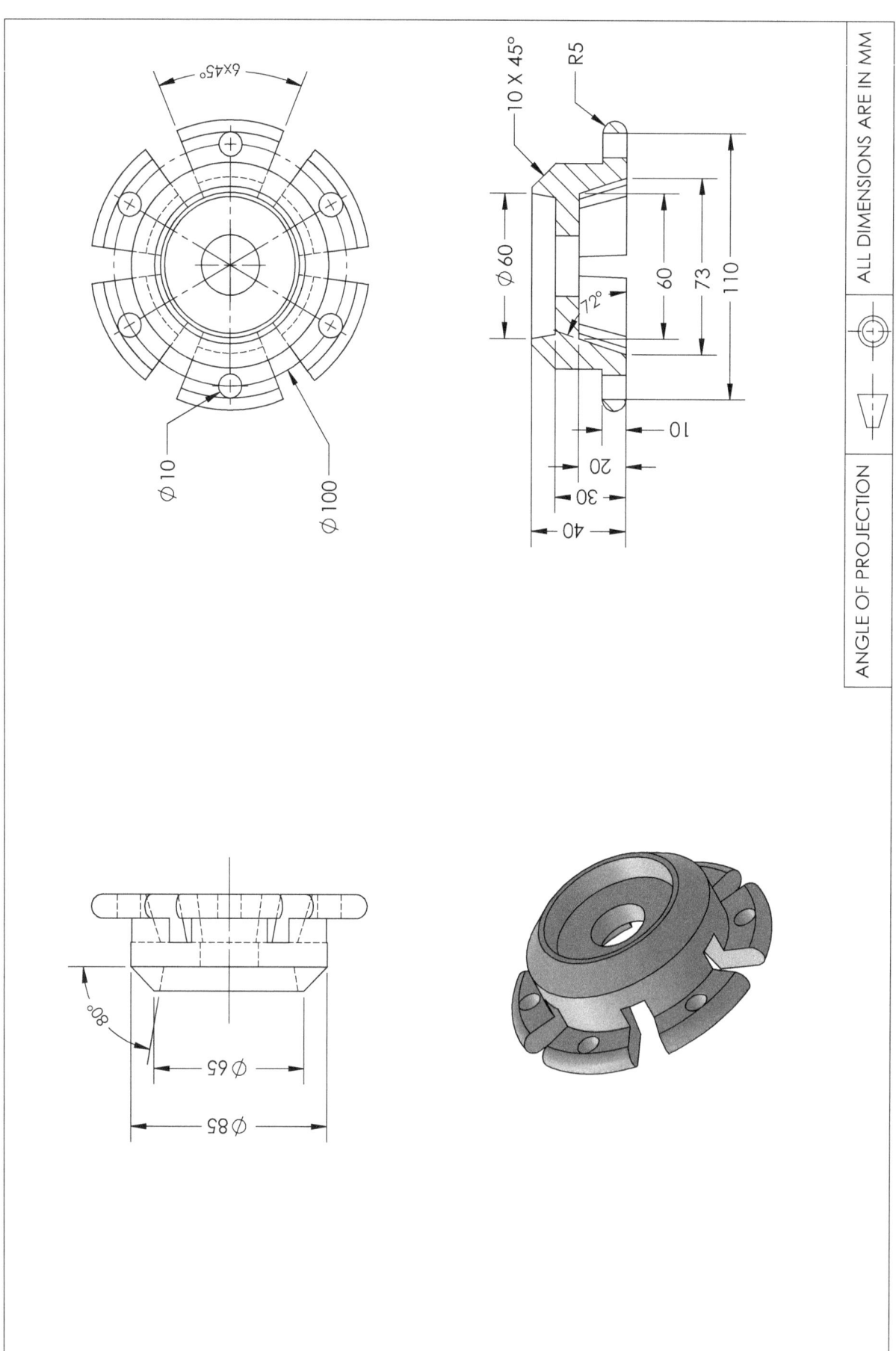

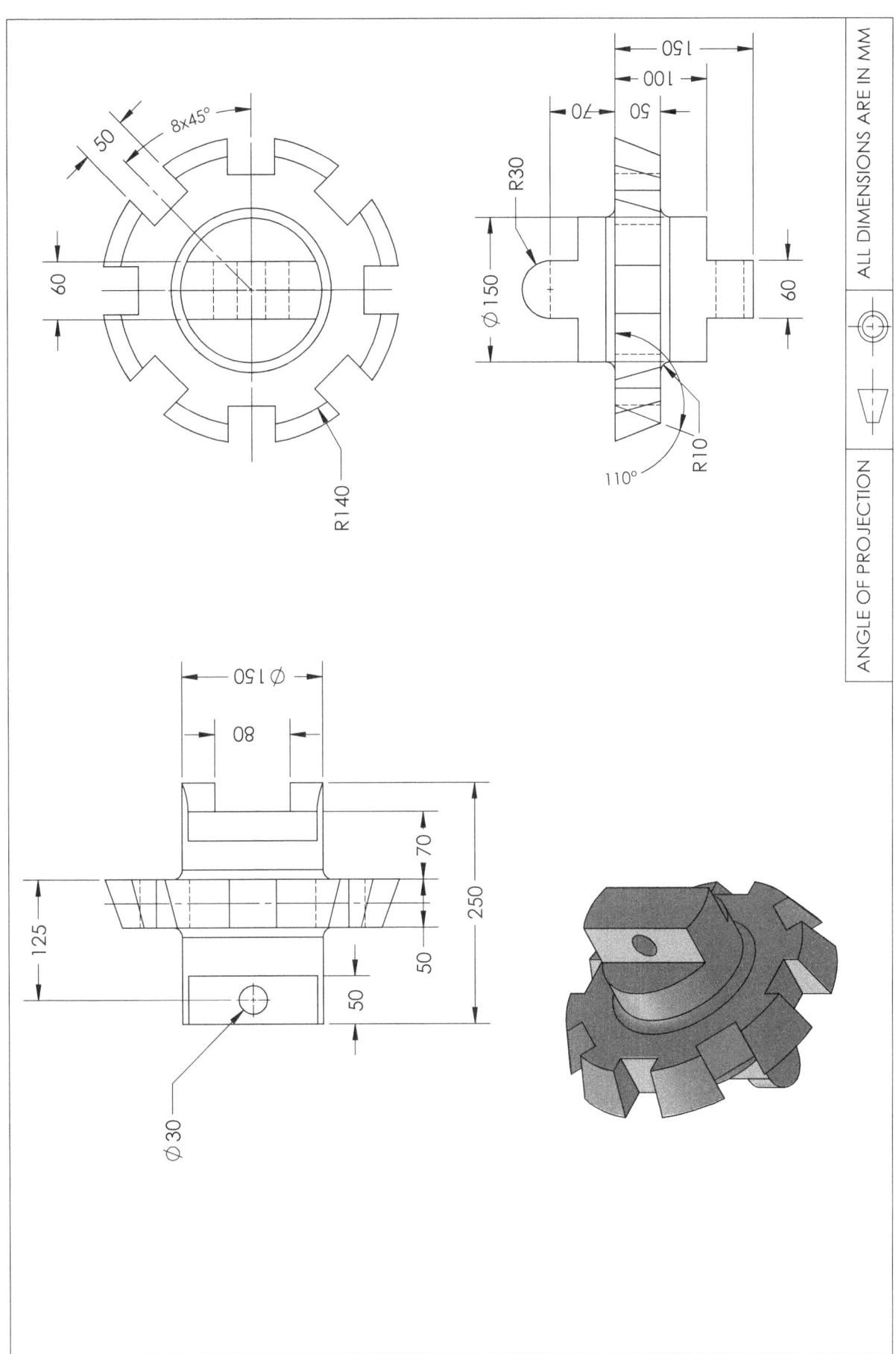

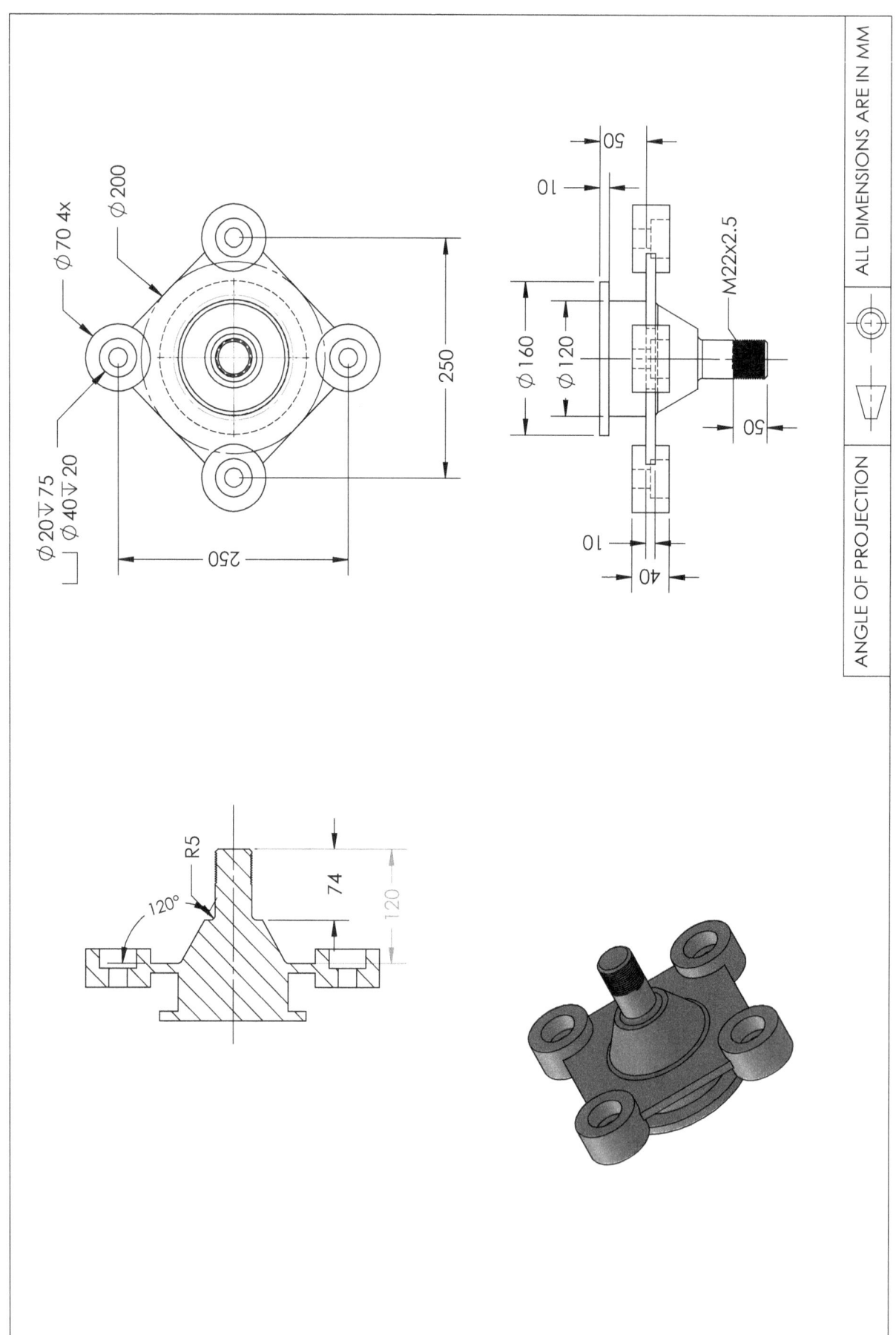

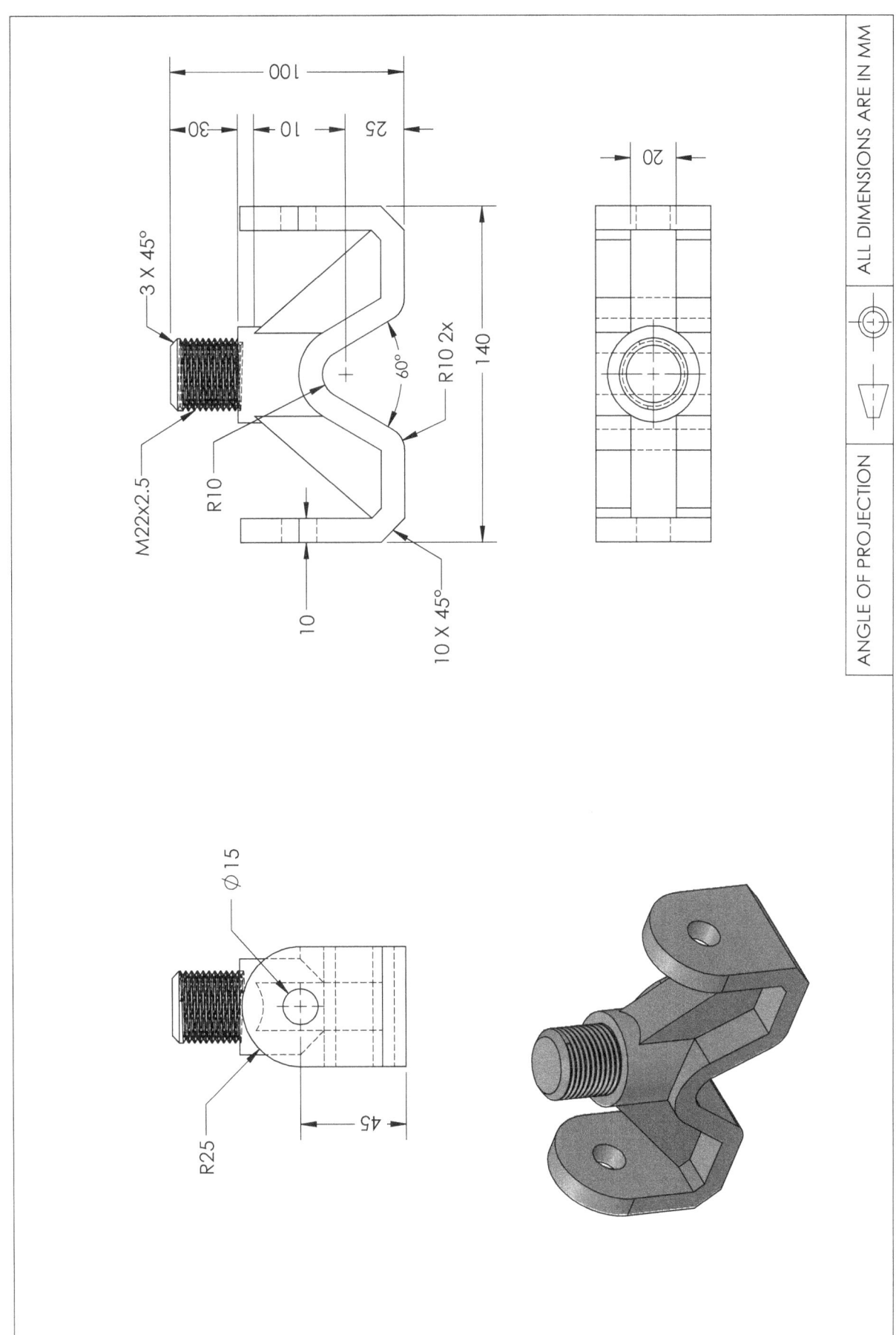

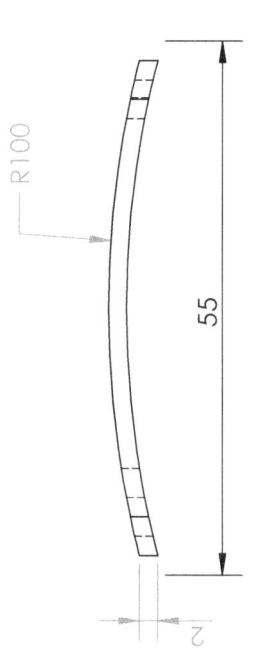
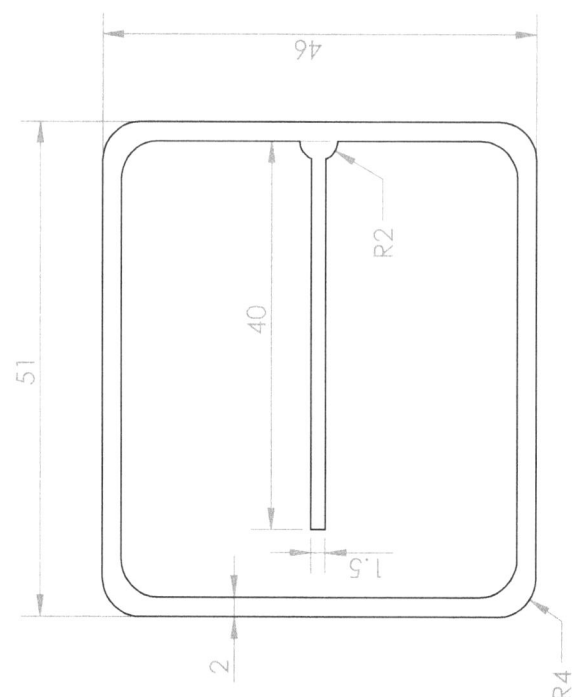
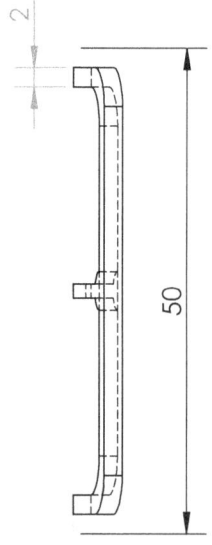
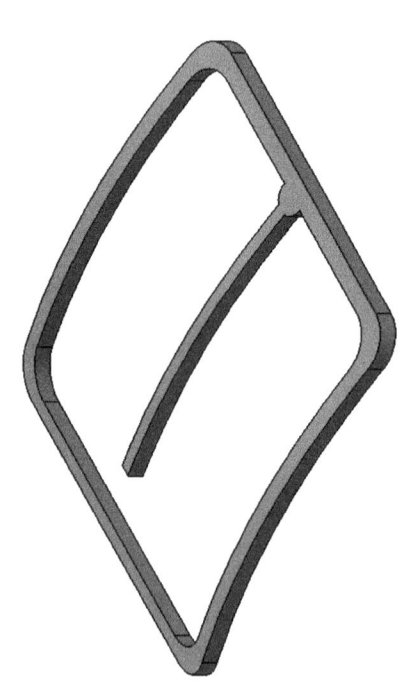

ALL DIMENSIONS ARE IN MM

ANGLE OF PROJECTION

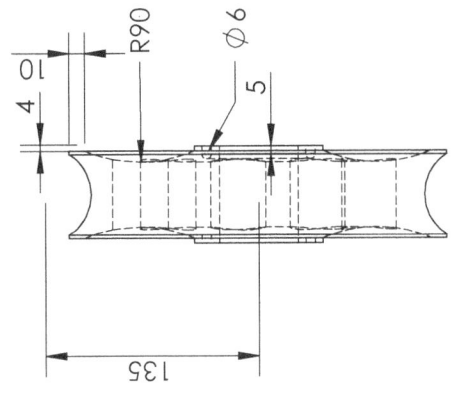
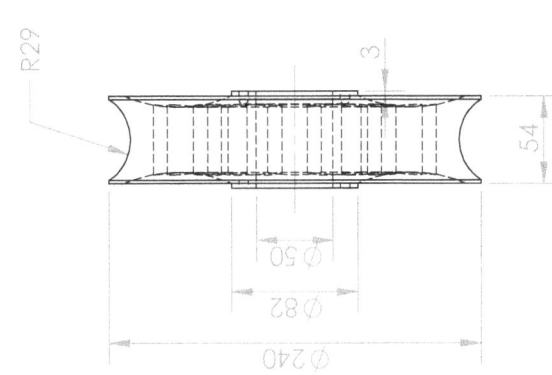
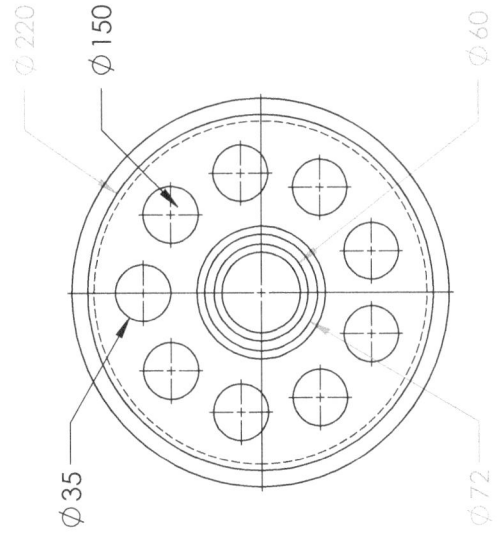
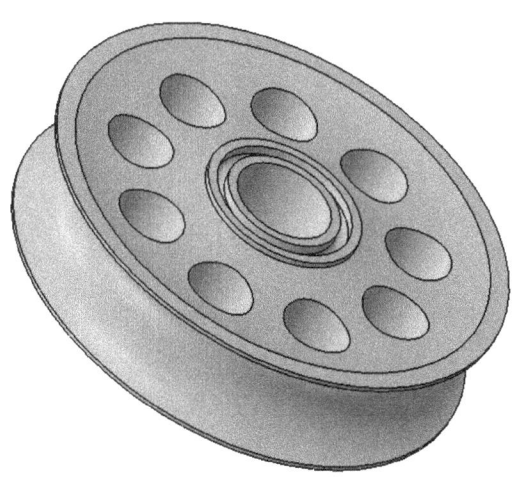

ALL DIMENSIONS ARE IN MM

ANGLE OF PROJECTION

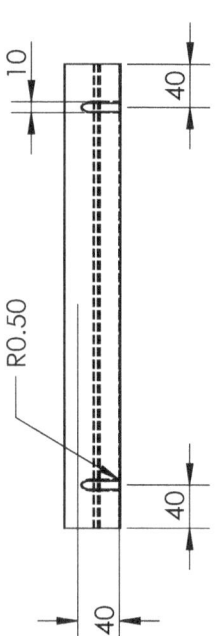
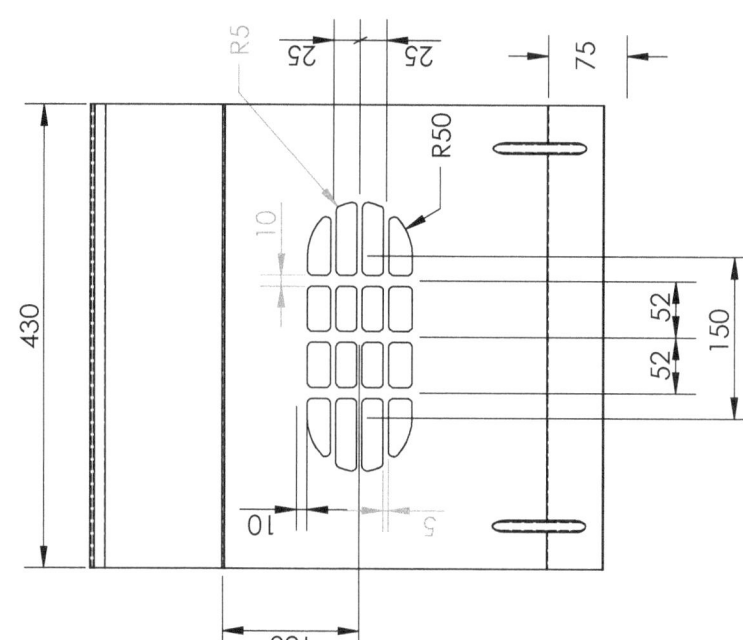
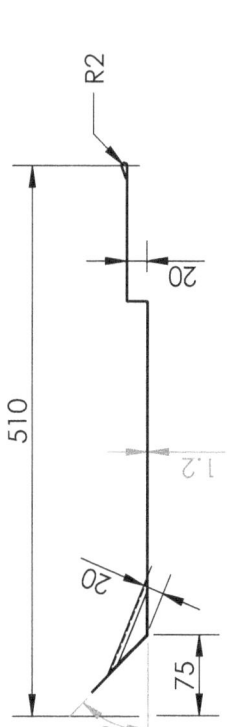
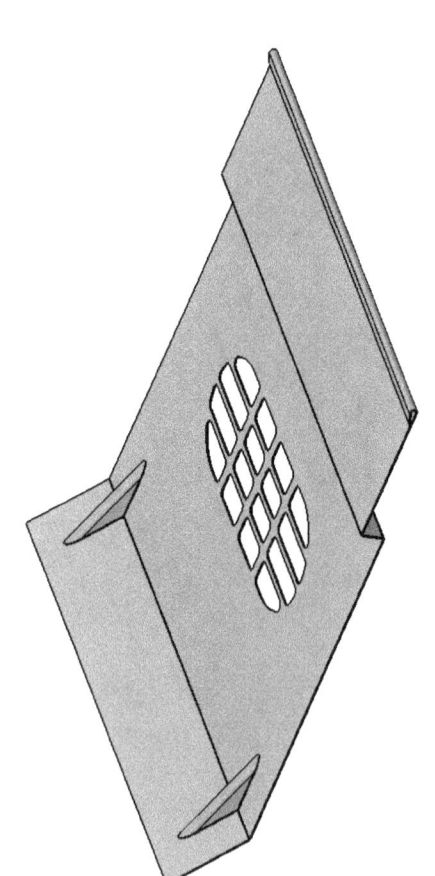

ALL DIMENSIONS ARE IN MM

ANGLE OF PROJECTION

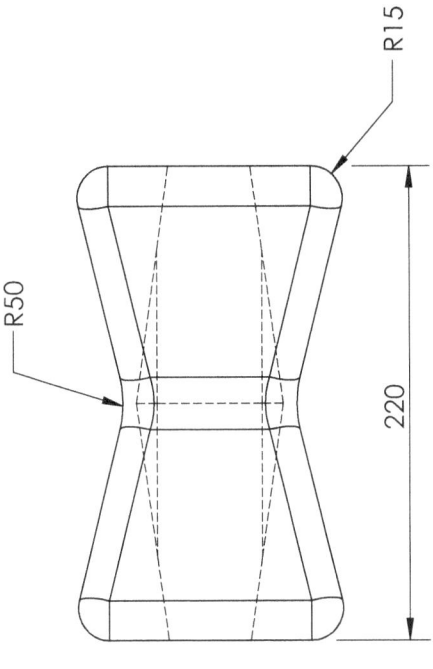
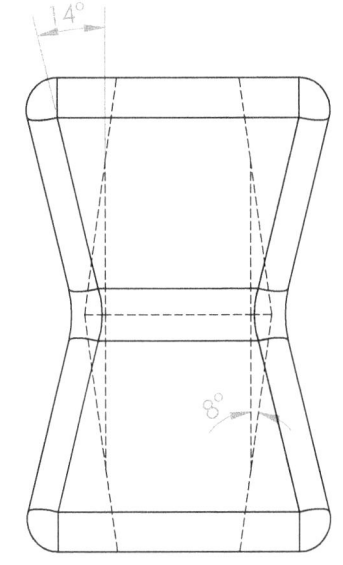
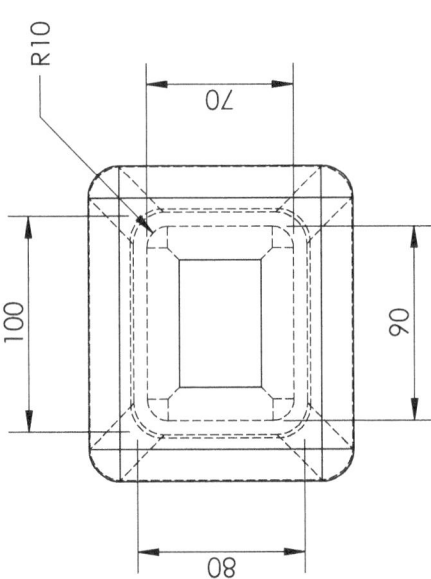
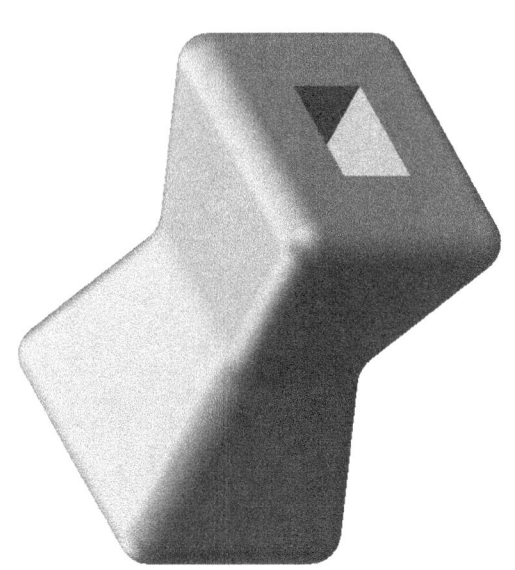

ALL DIMENSIONS ARE IN MM

ANGLE OF PROJECTION

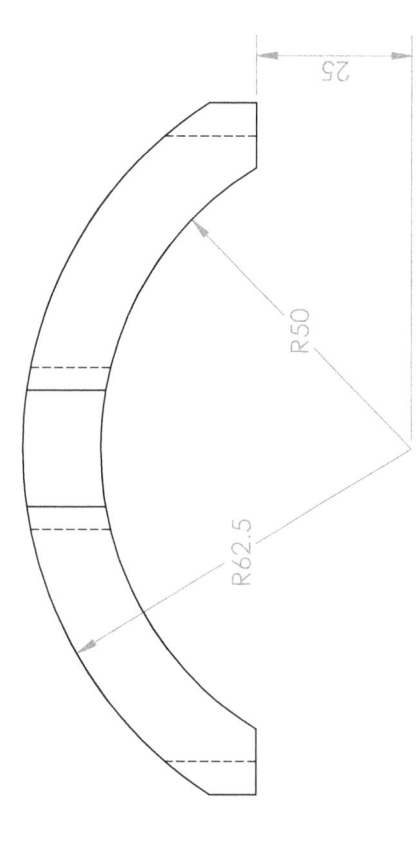
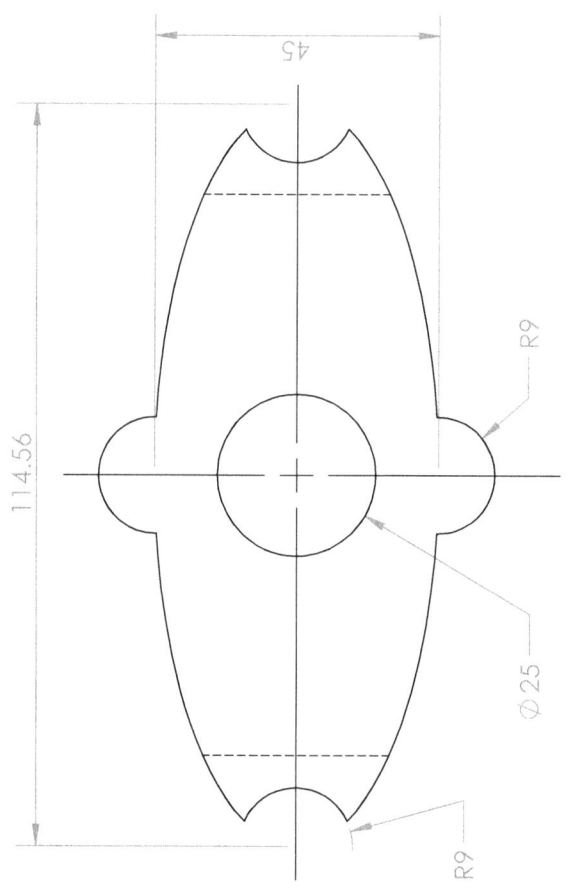
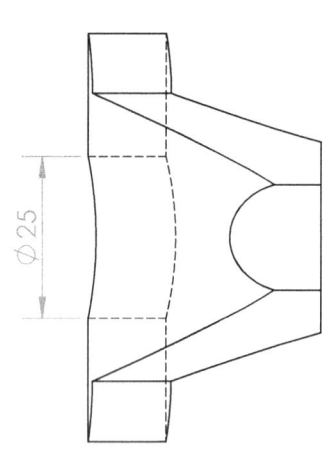
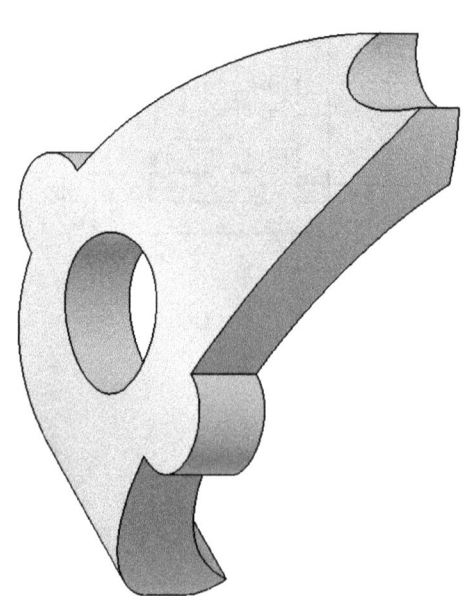

ALL DIMENSIONS ARE IN MM

ANGLE OF PROJECTION

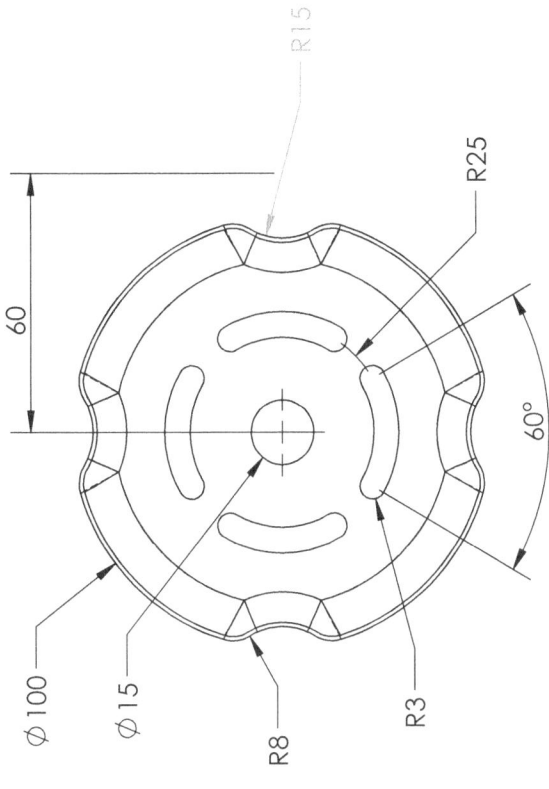
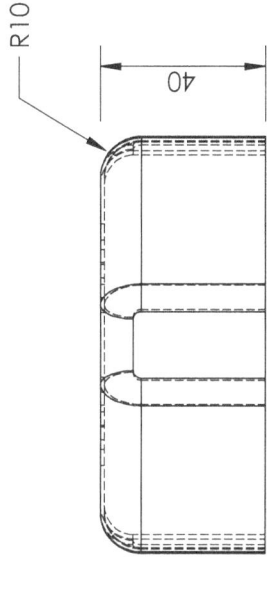
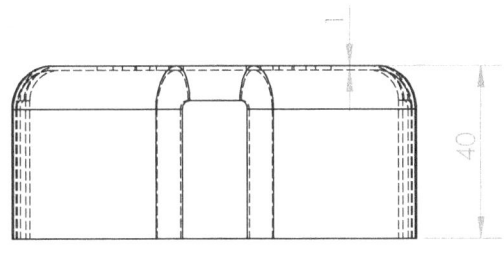
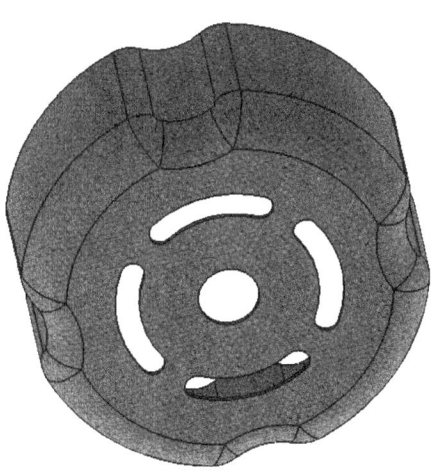

ALL DIMENSIONS ARE IN MM

ANGLE OF PROJECTION

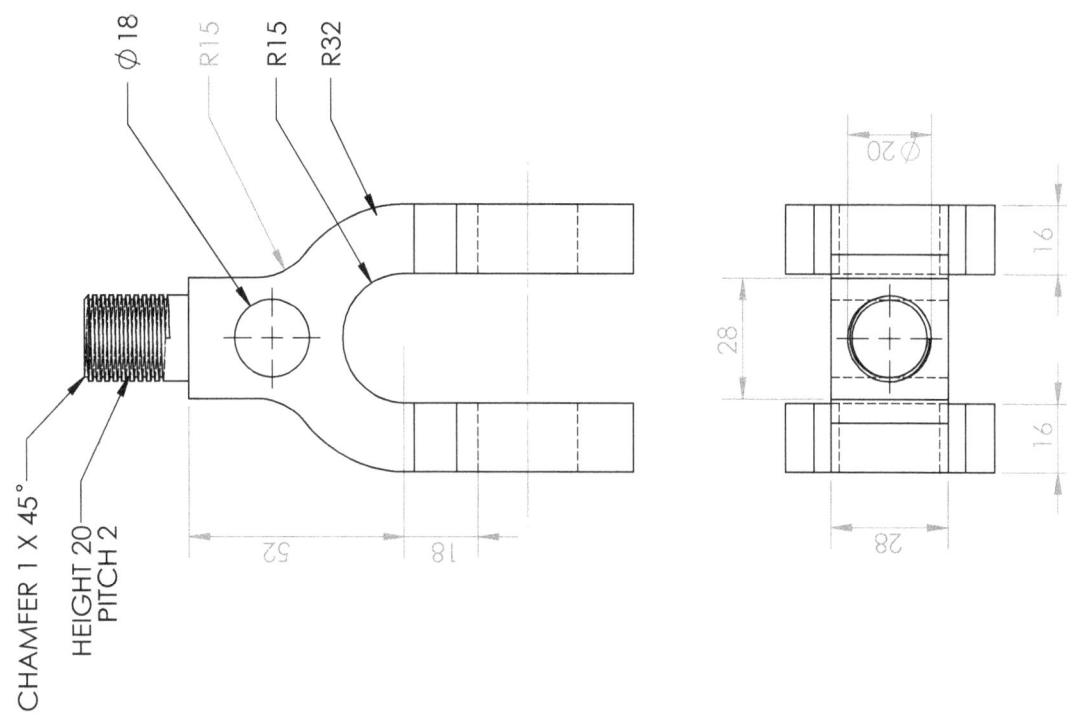
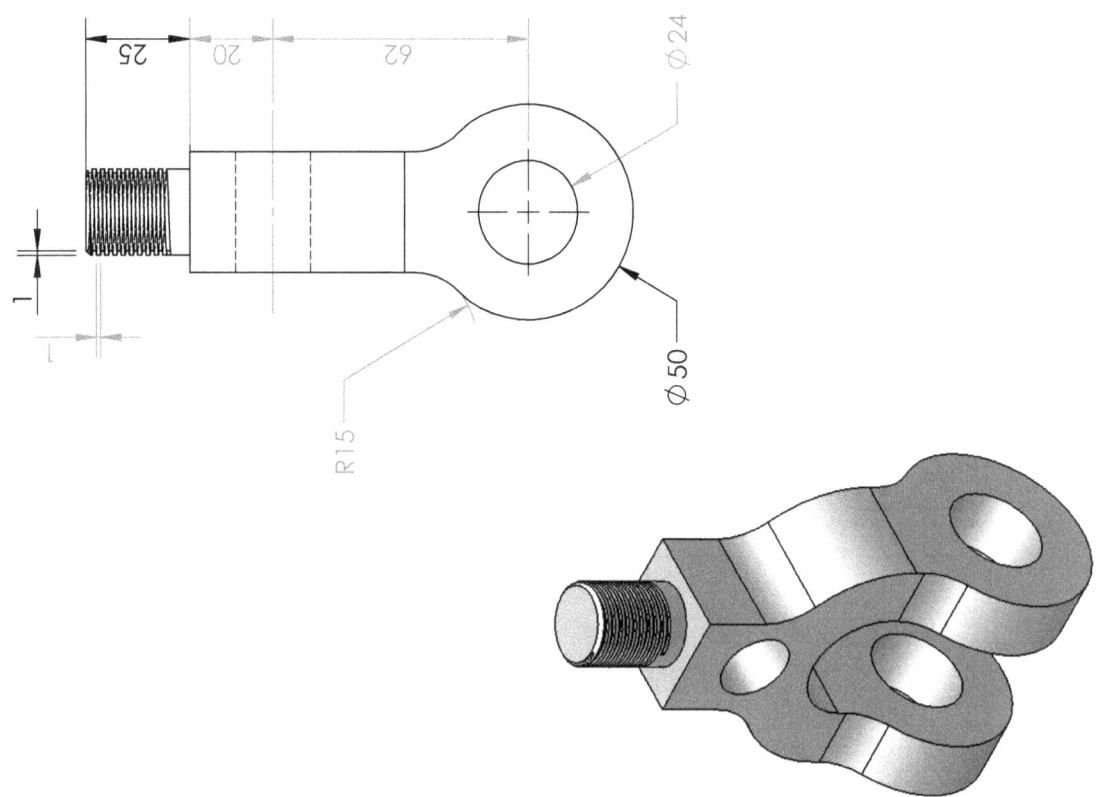

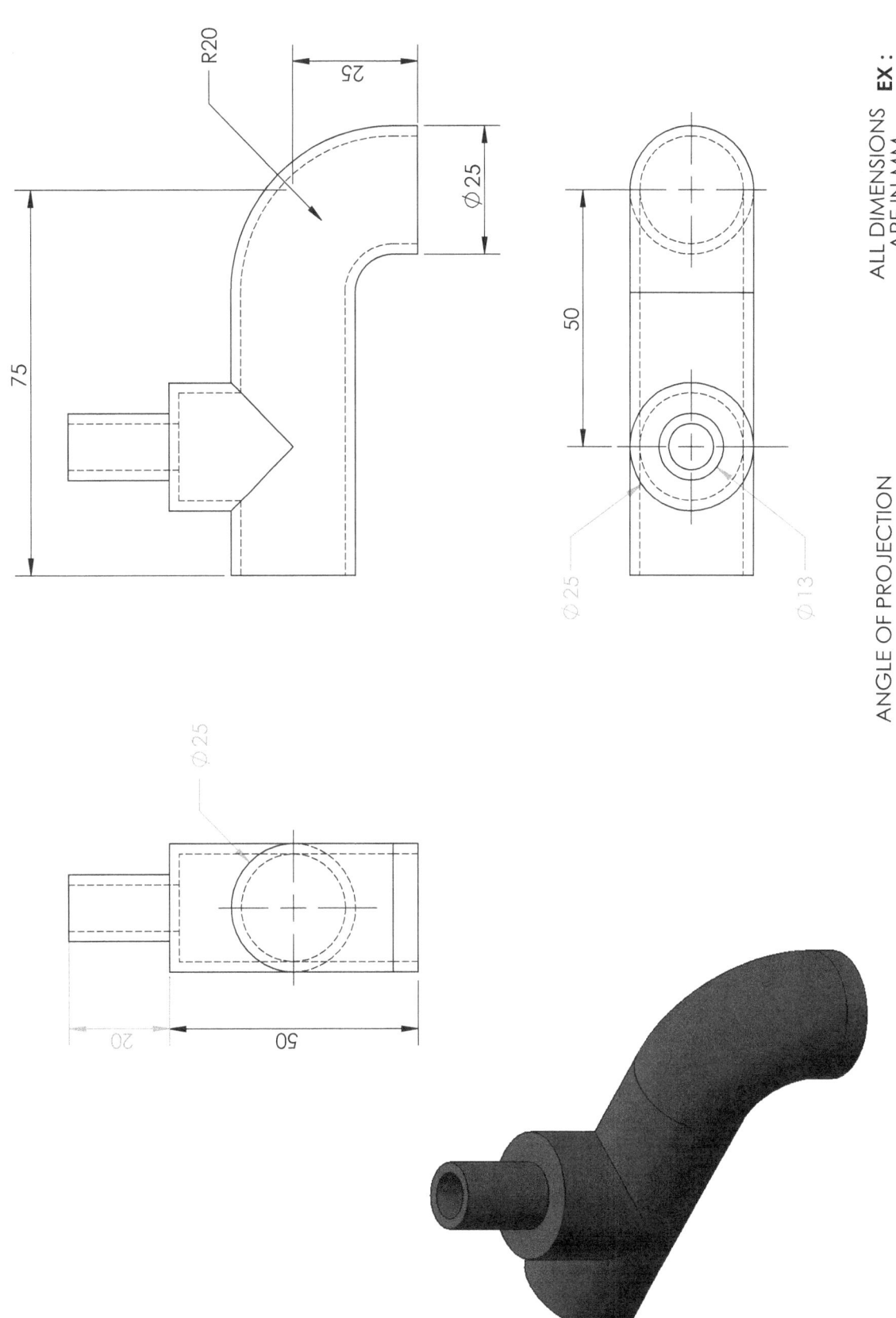

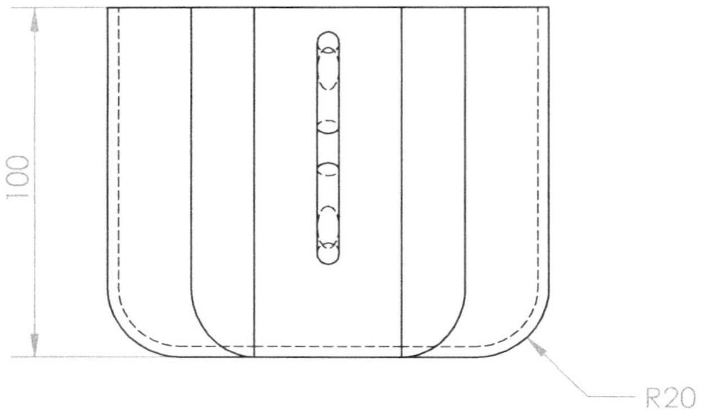

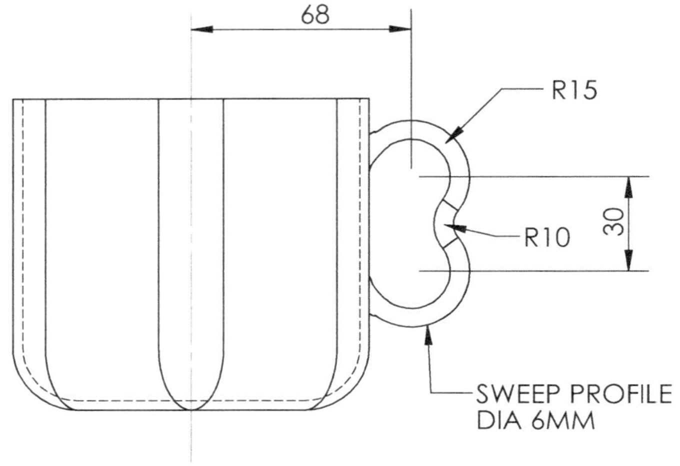

R15

R10

30

SWEEP PROFILE
DIA 6MM

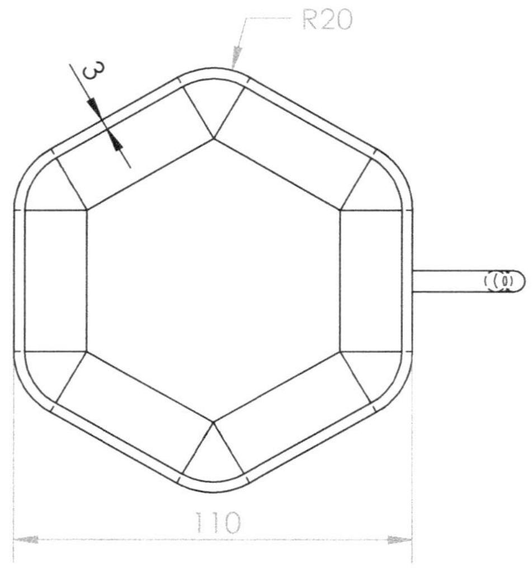

68

R20

110

3

R20

100

ALL DIMENSIONS ARE IN MM

EX :

ANGLE OF PROJECTION

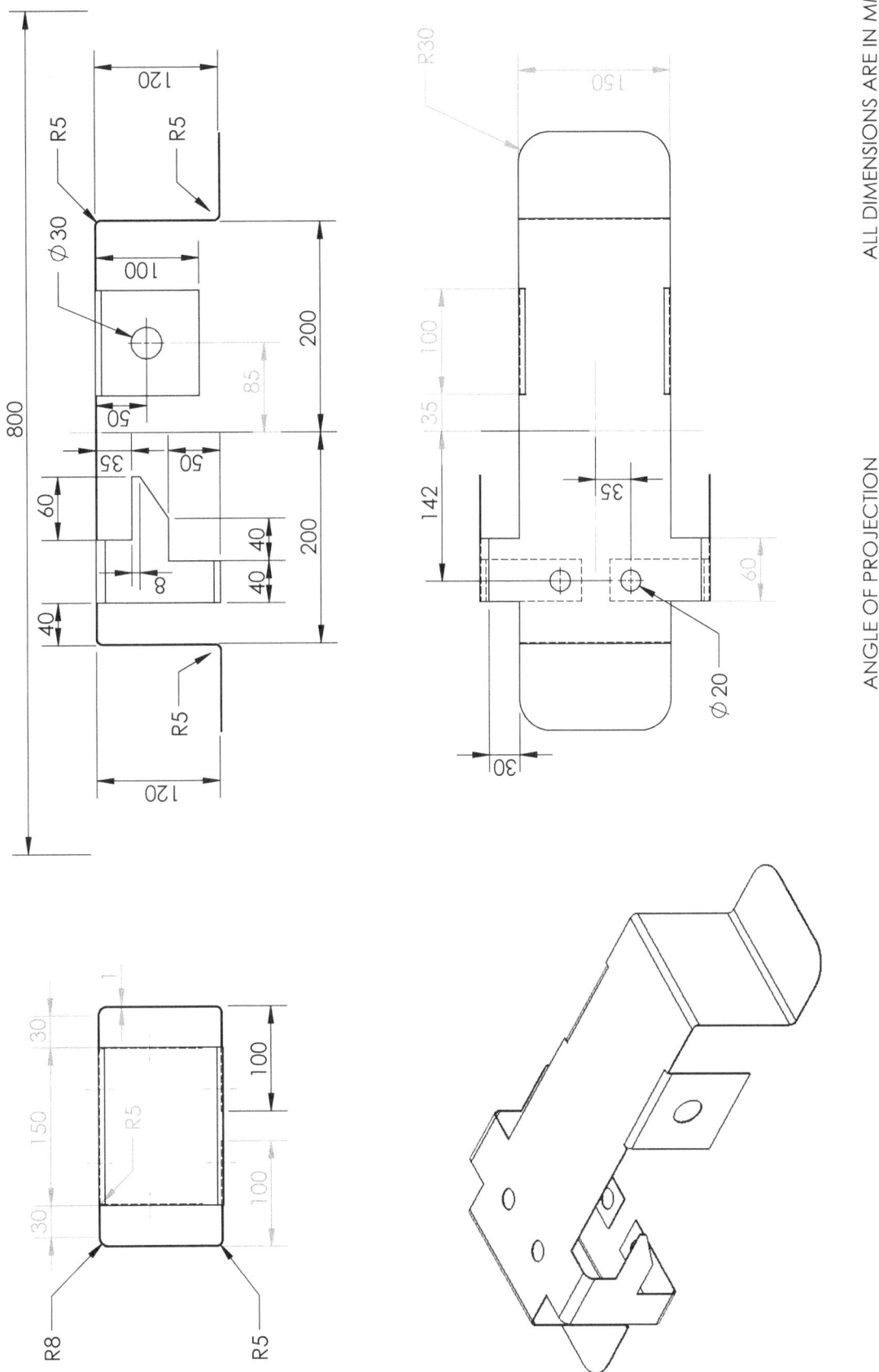

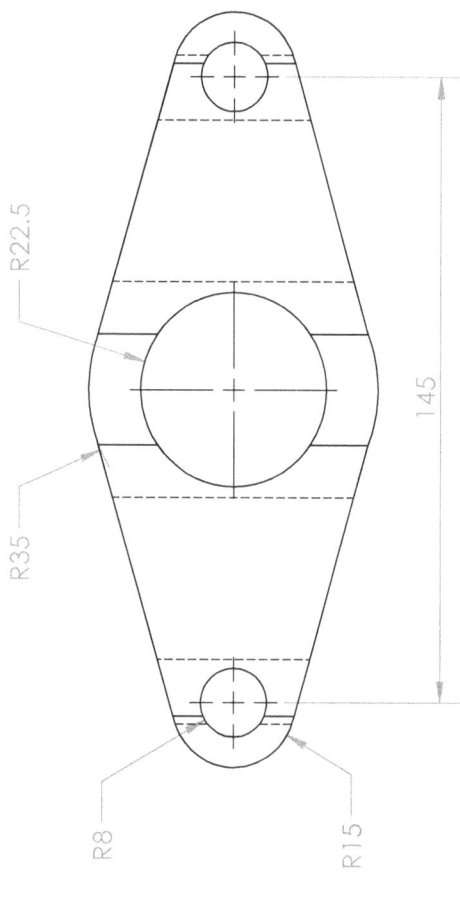
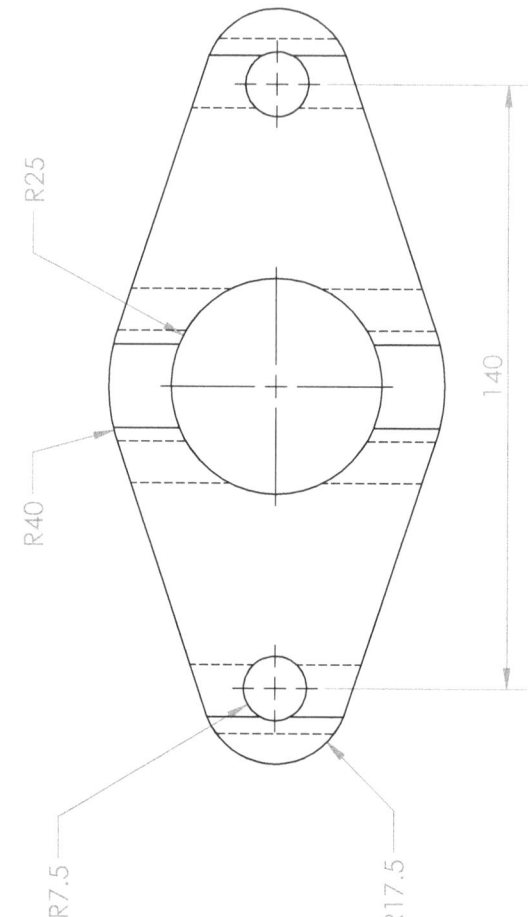

ALL DIMENSIONS ARE IN MM

ANGLE OF PROJECTION

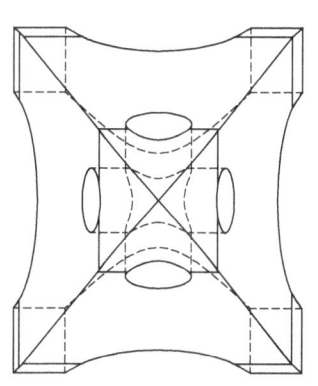
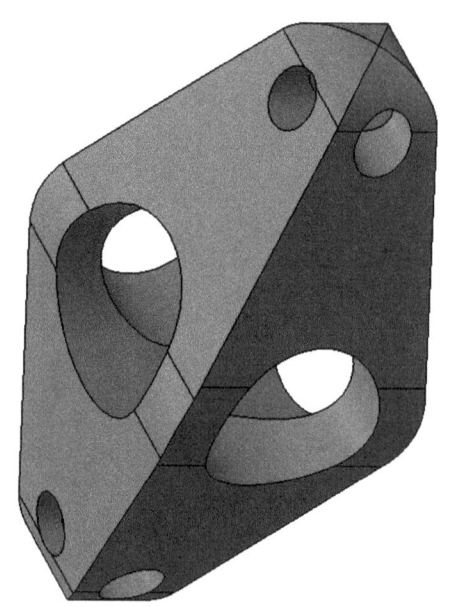

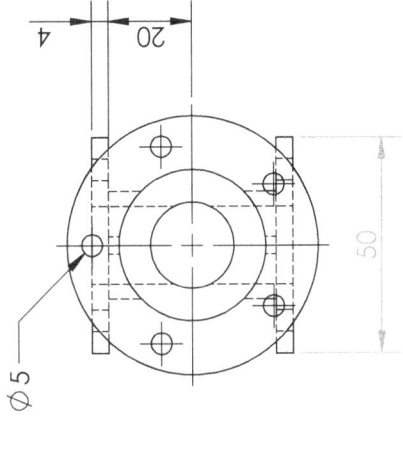
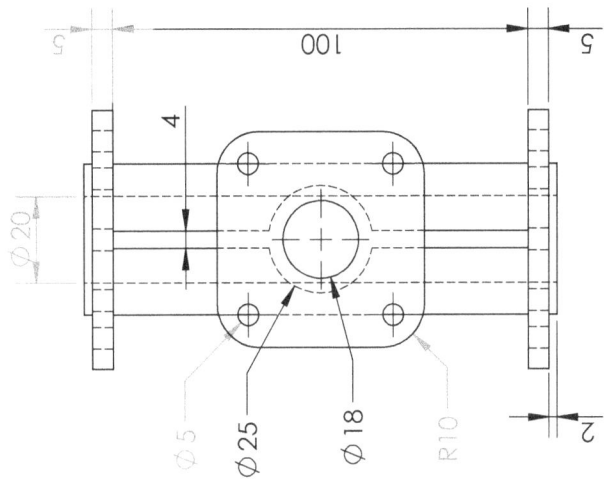
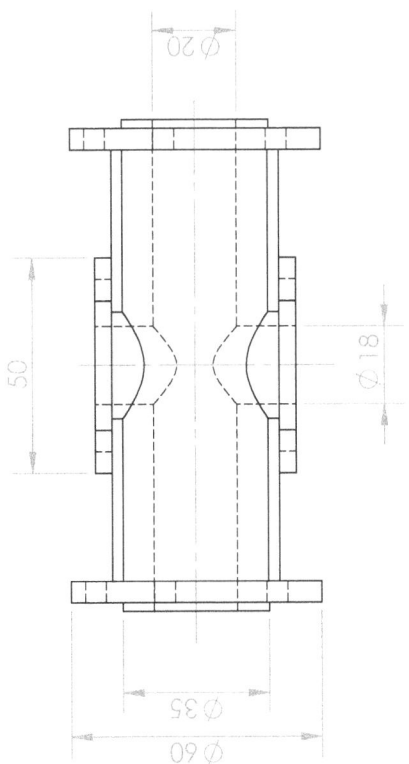
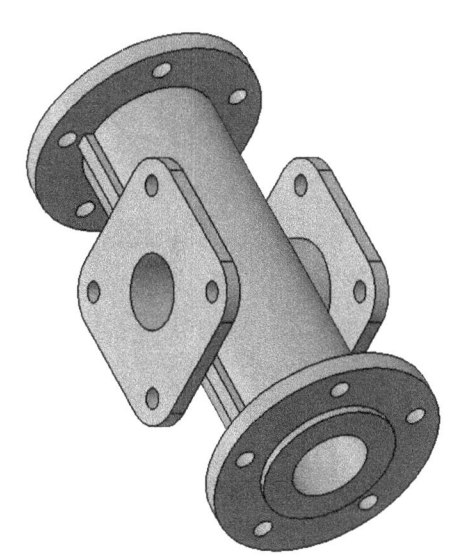

ALL DIMENSIONS ARE IN MM

ANGLE OF PROJECTION

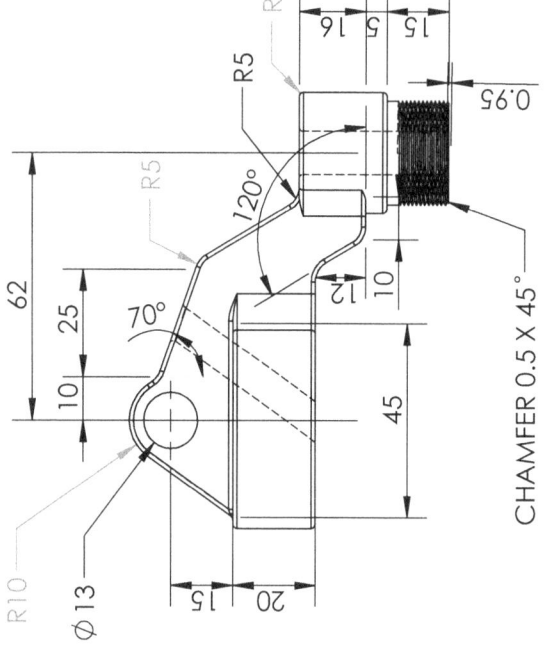
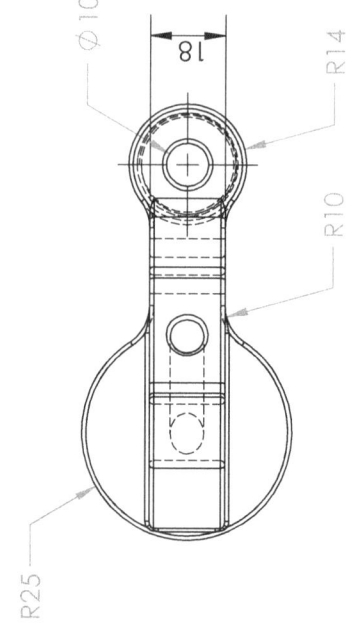
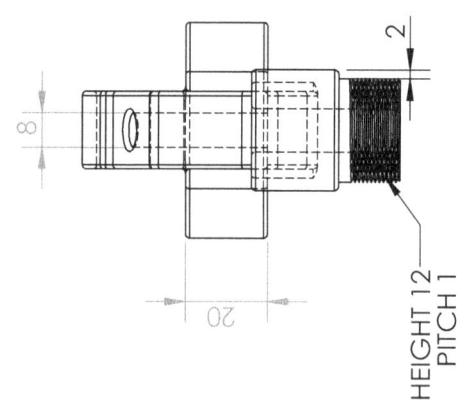
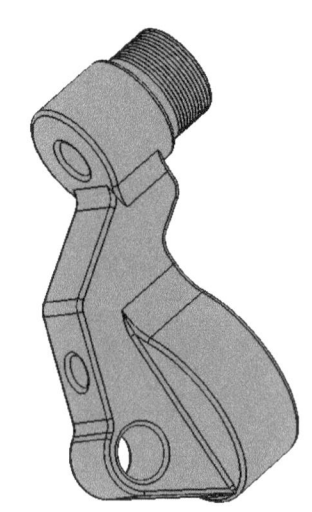

ALL DIMENSIONS ARE IN MM

ANGLE OF PROJECTION

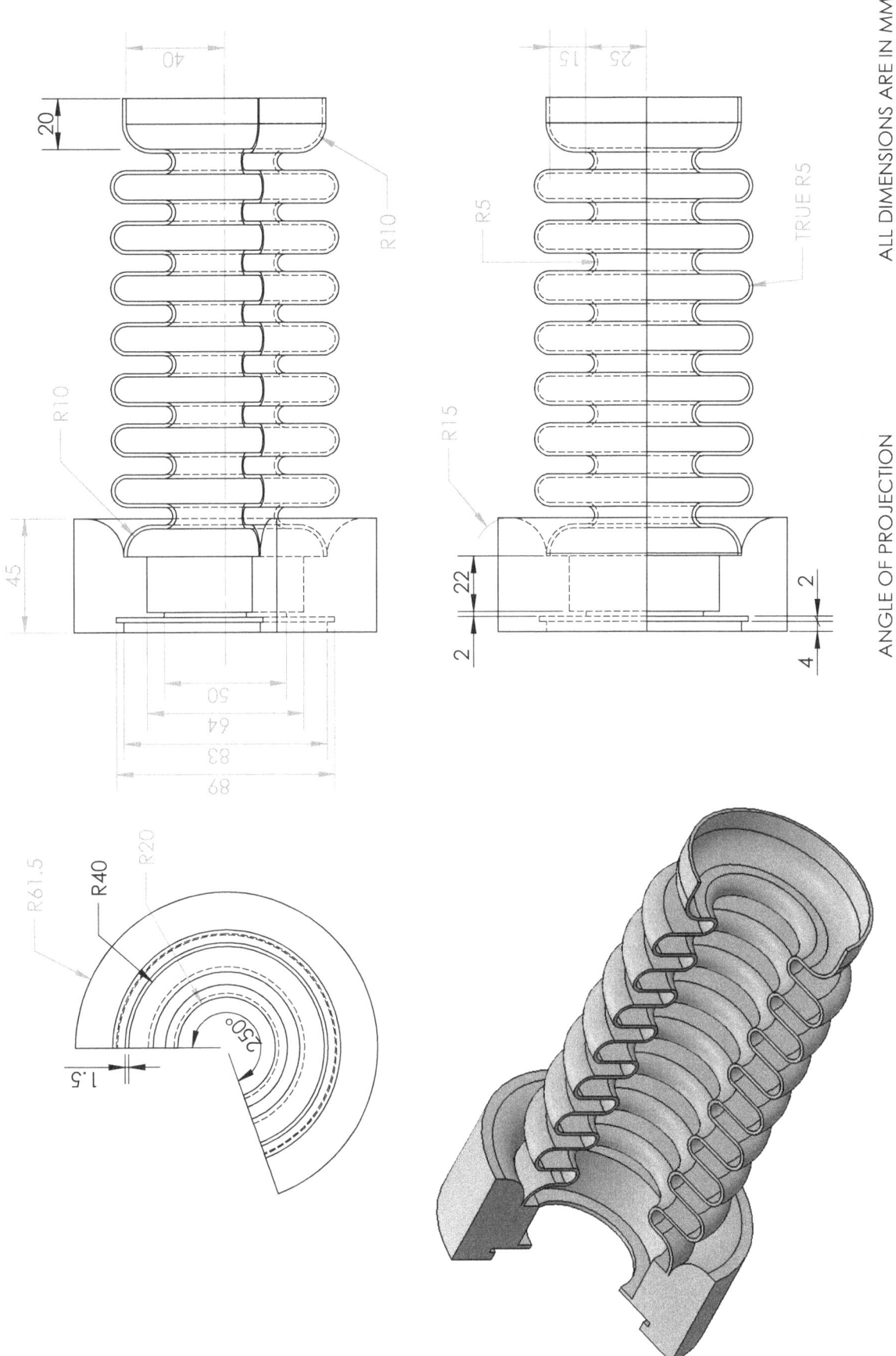

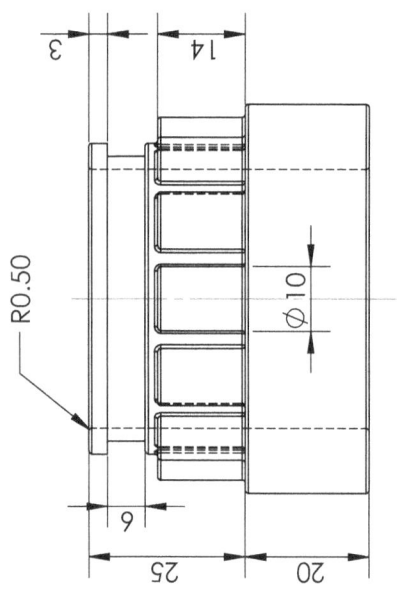
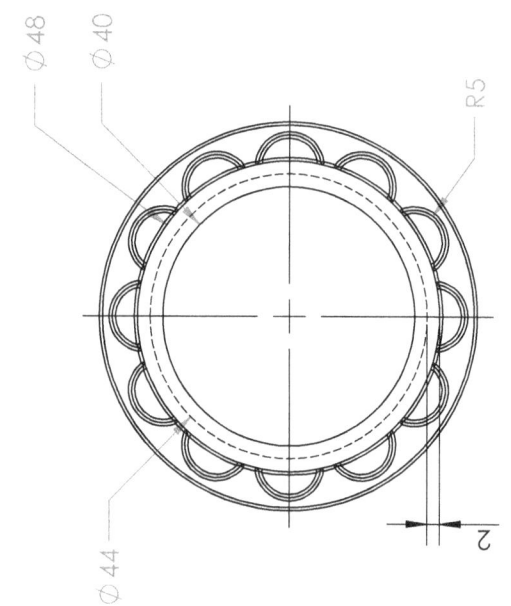
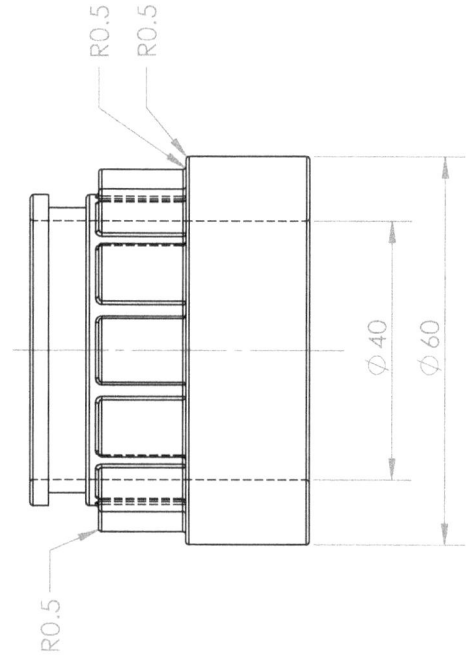
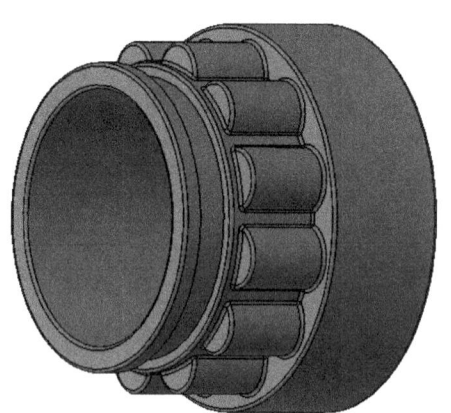

ALL DIMENSIONS ARE IN MM

ANGLE OF PROJECTION

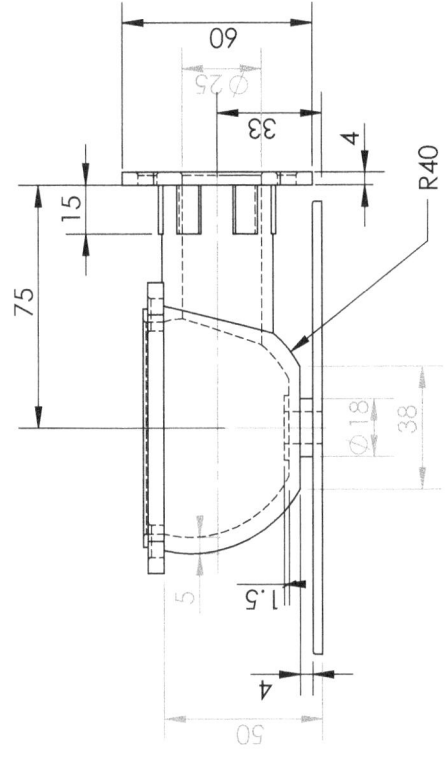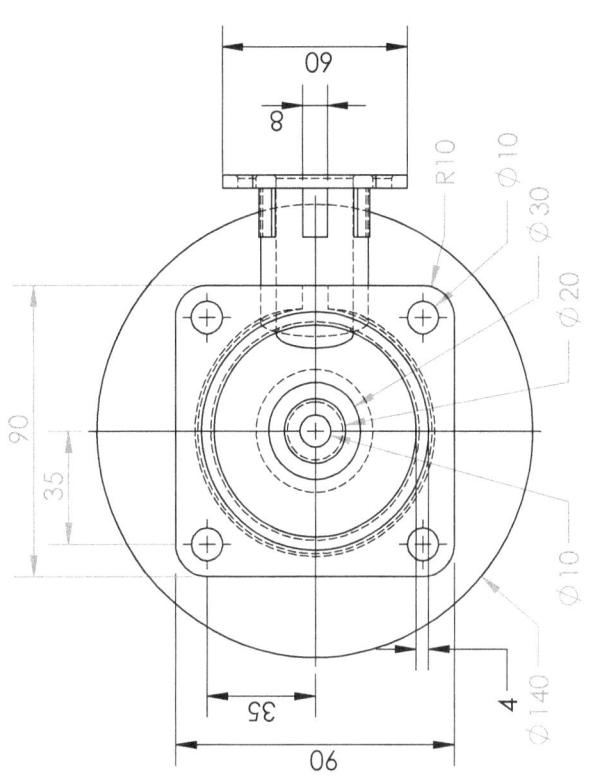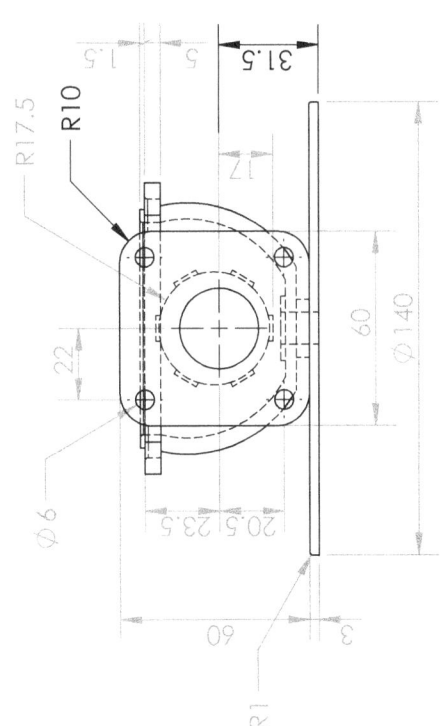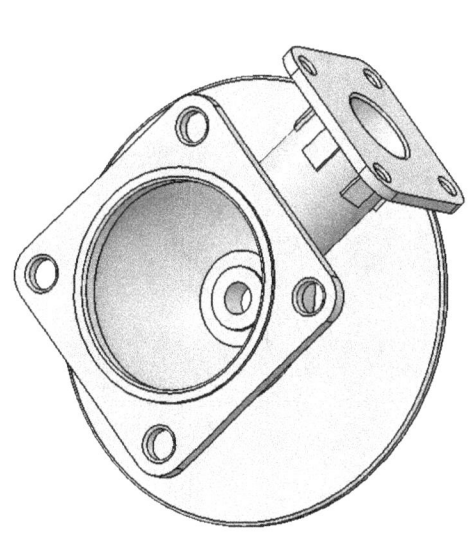

ALL DIMENSIONS ARE IN MM

ANGLE OF PROJECTION

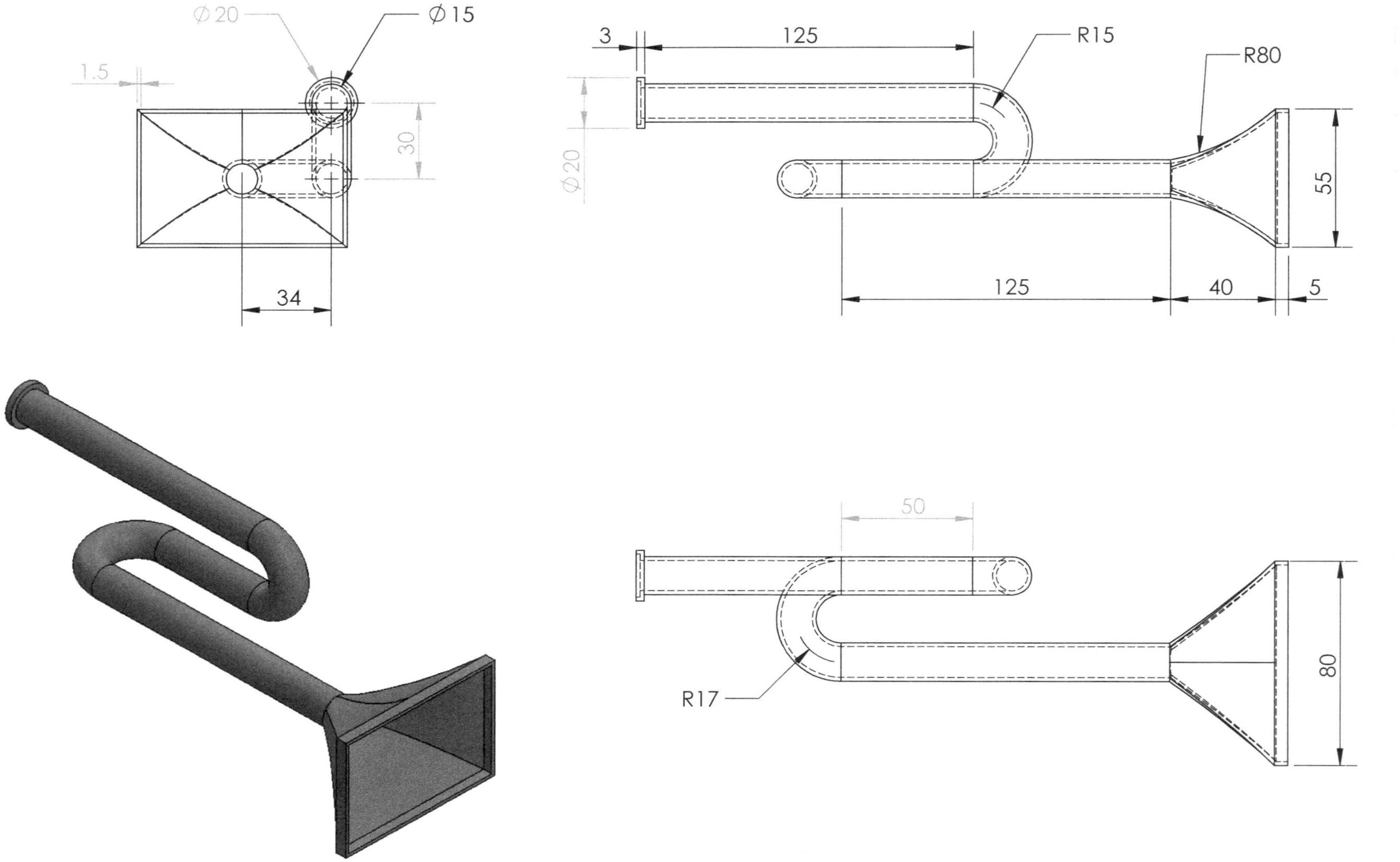

ALL DIMENSIONS ARE IN MM

ANGLE OF PROJECTION

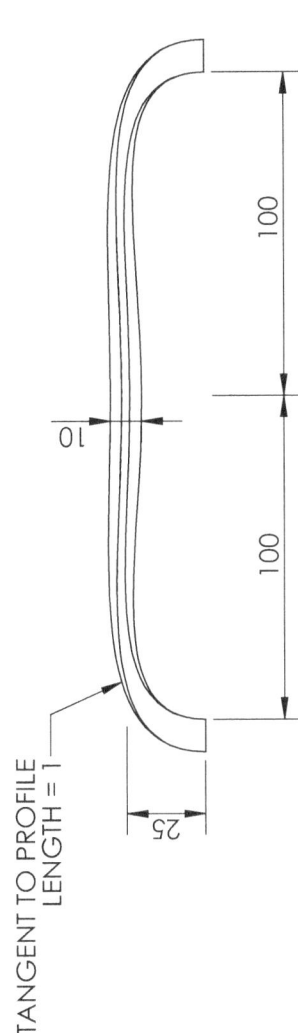

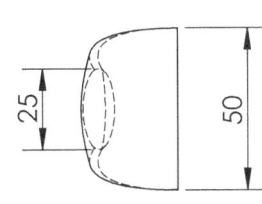

TANGENT TO PROFILE
LENGTH = 1

ALL DIMENSIONS ARE IN MM

ANGLE OF PROJECTION

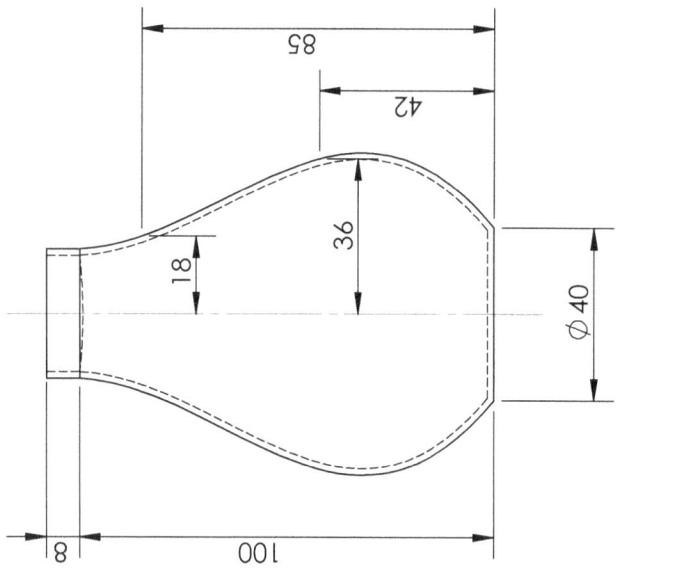
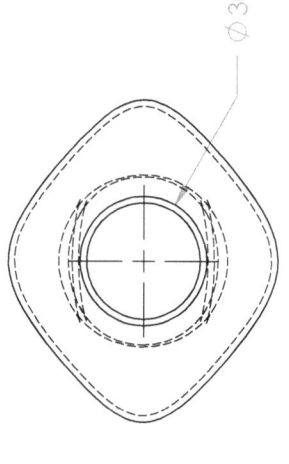
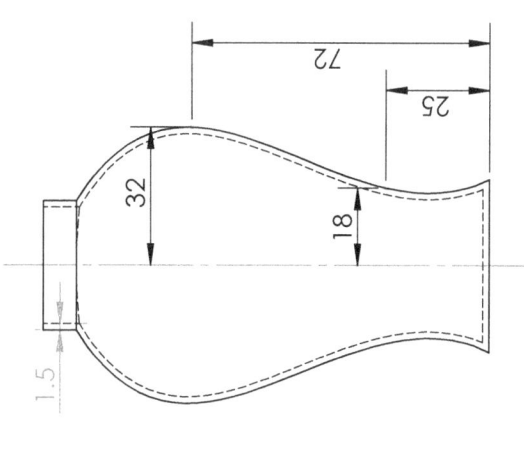
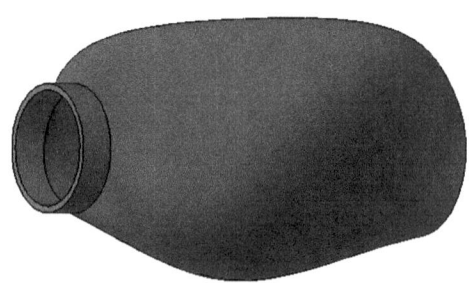

ALL DIMENSIONS ARE IN MM

ANGLE OF PROJECTION   EX :

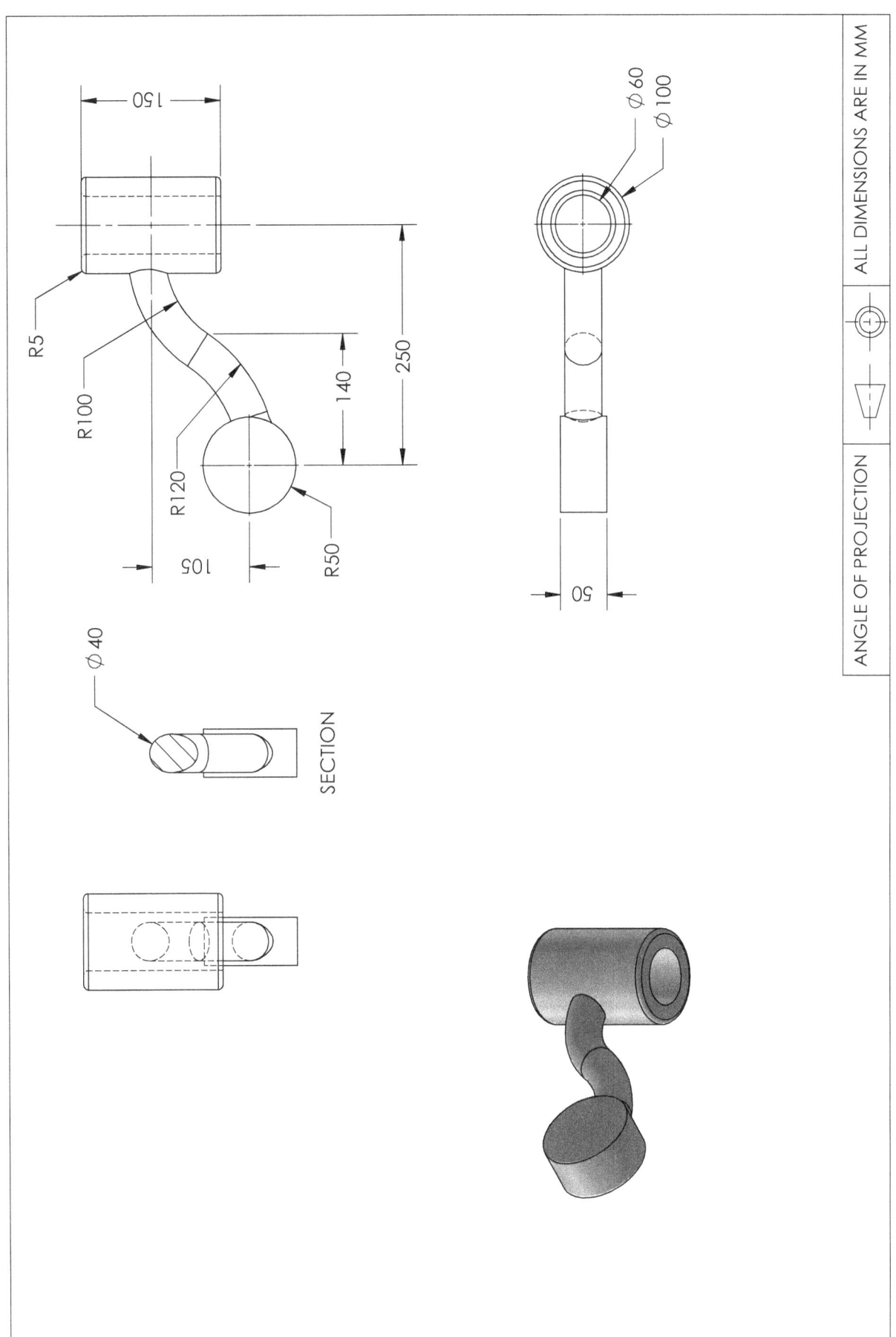

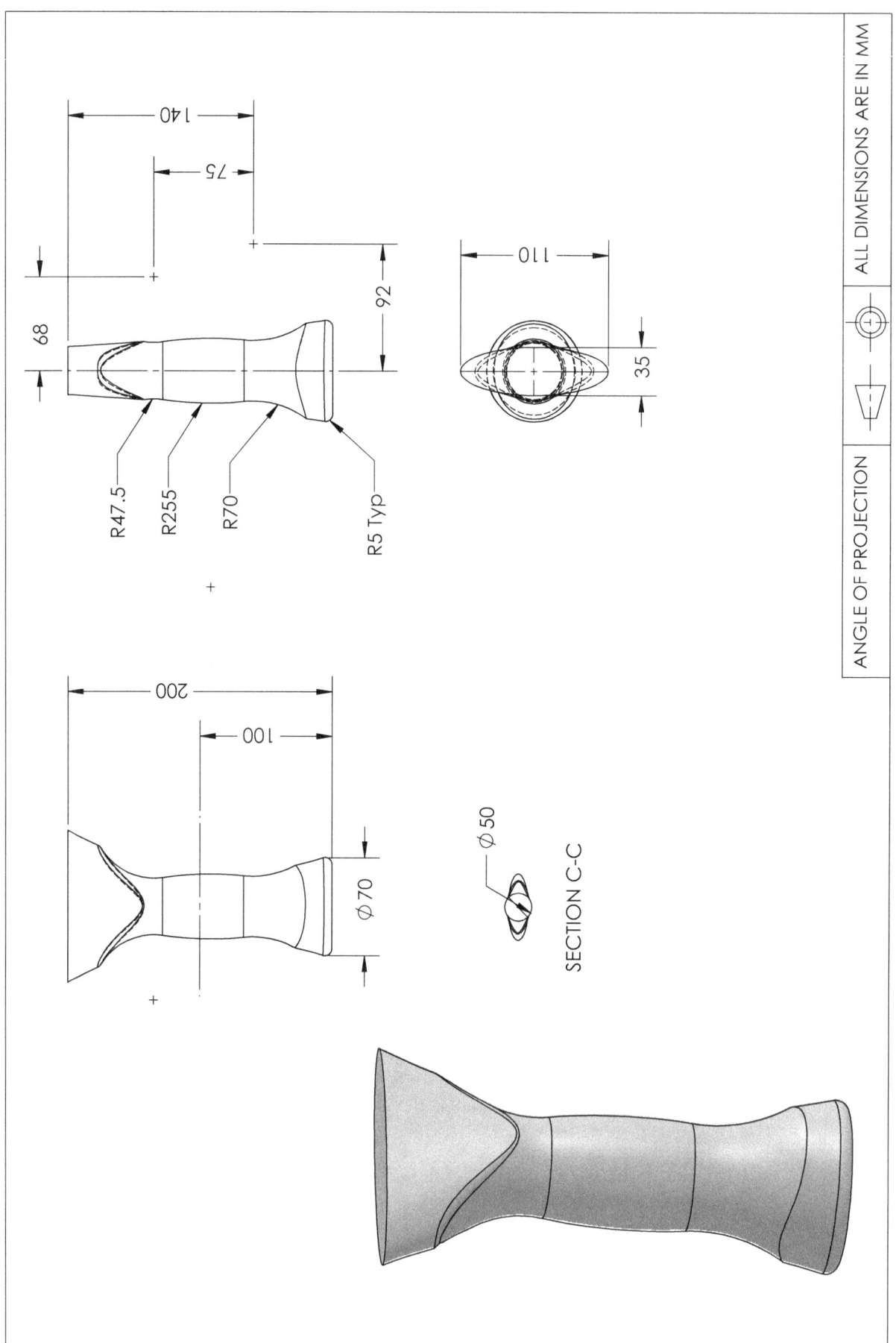

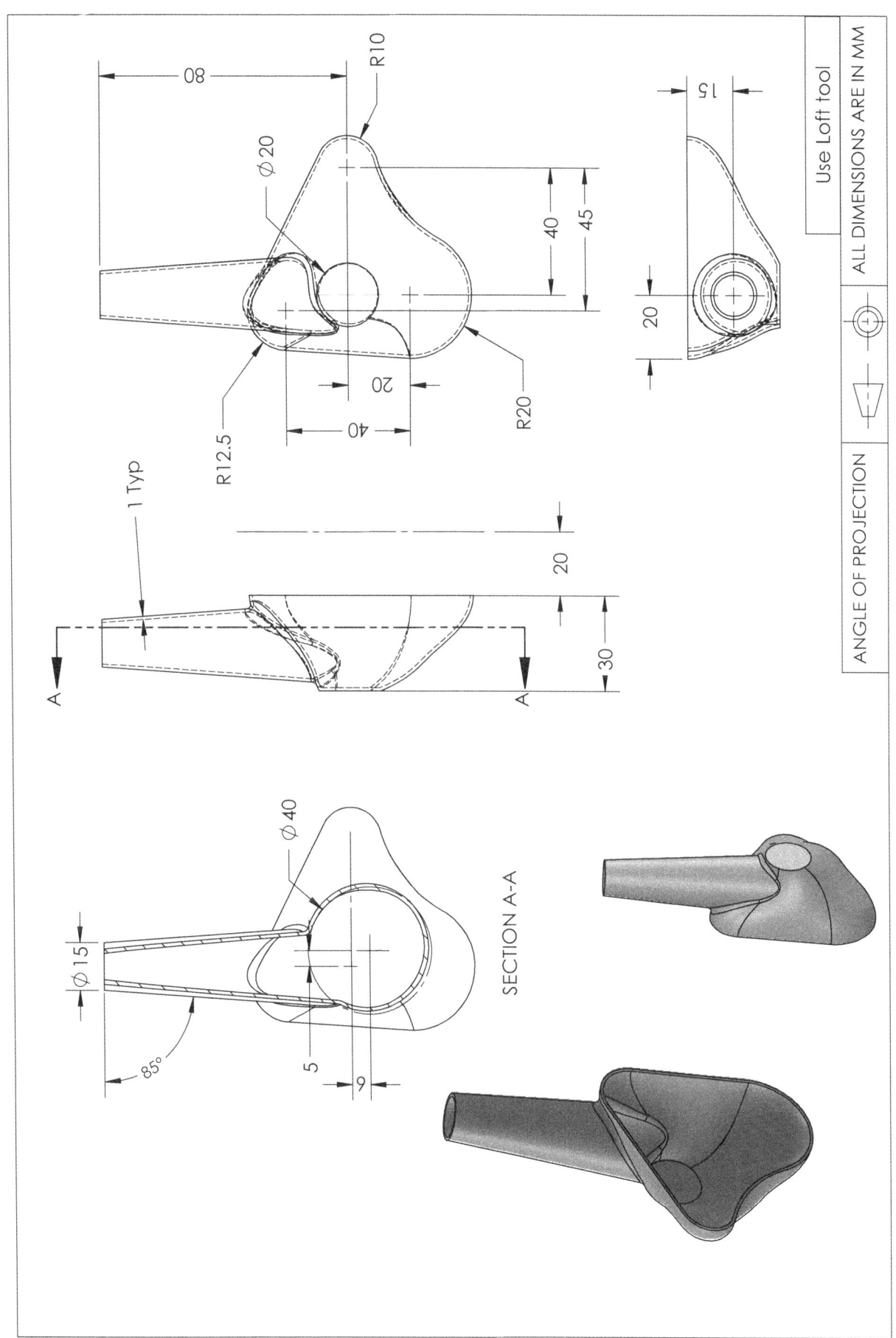

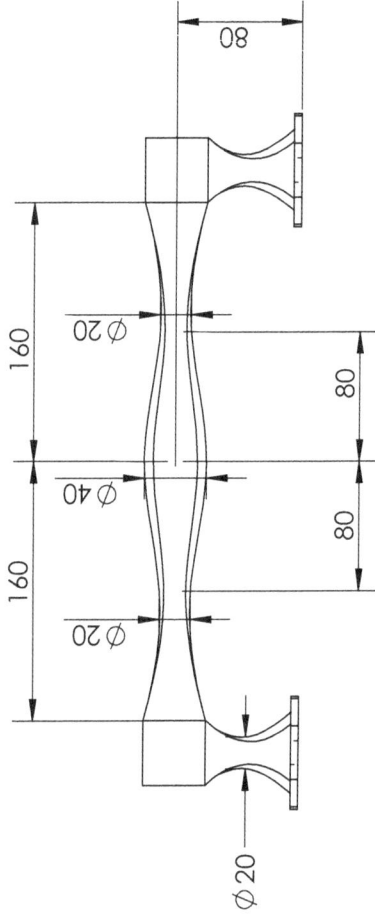
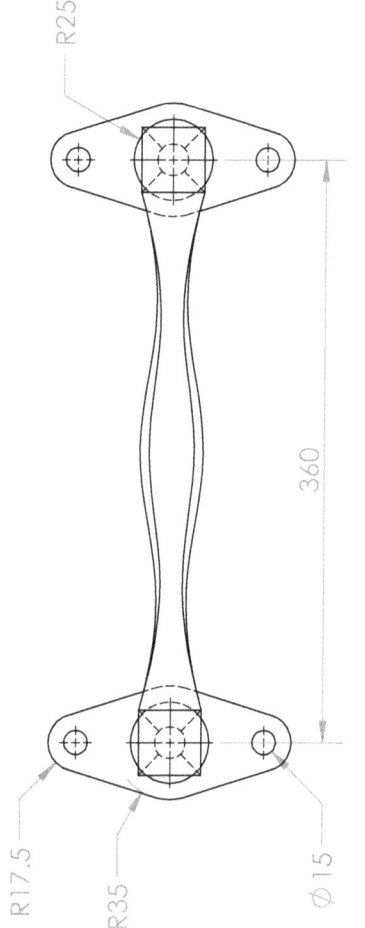
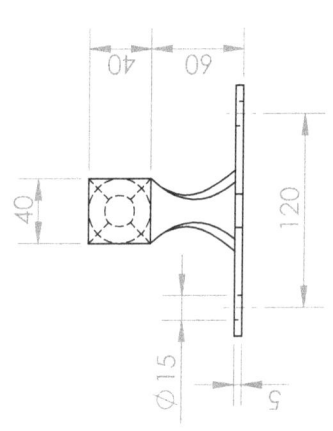
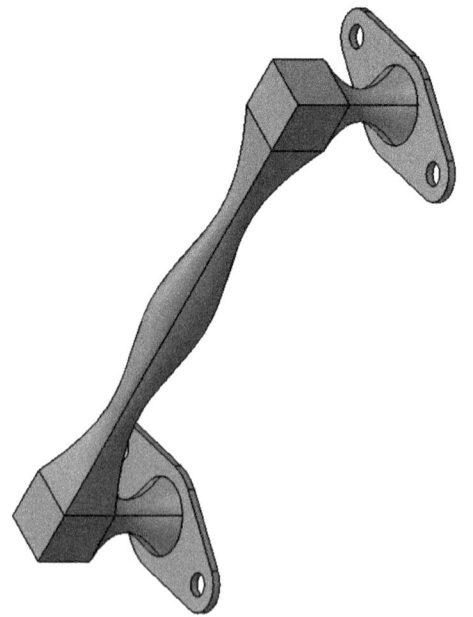

ALL DIMENSIONS ARE IN MM

ANGLE OF PROJECTION

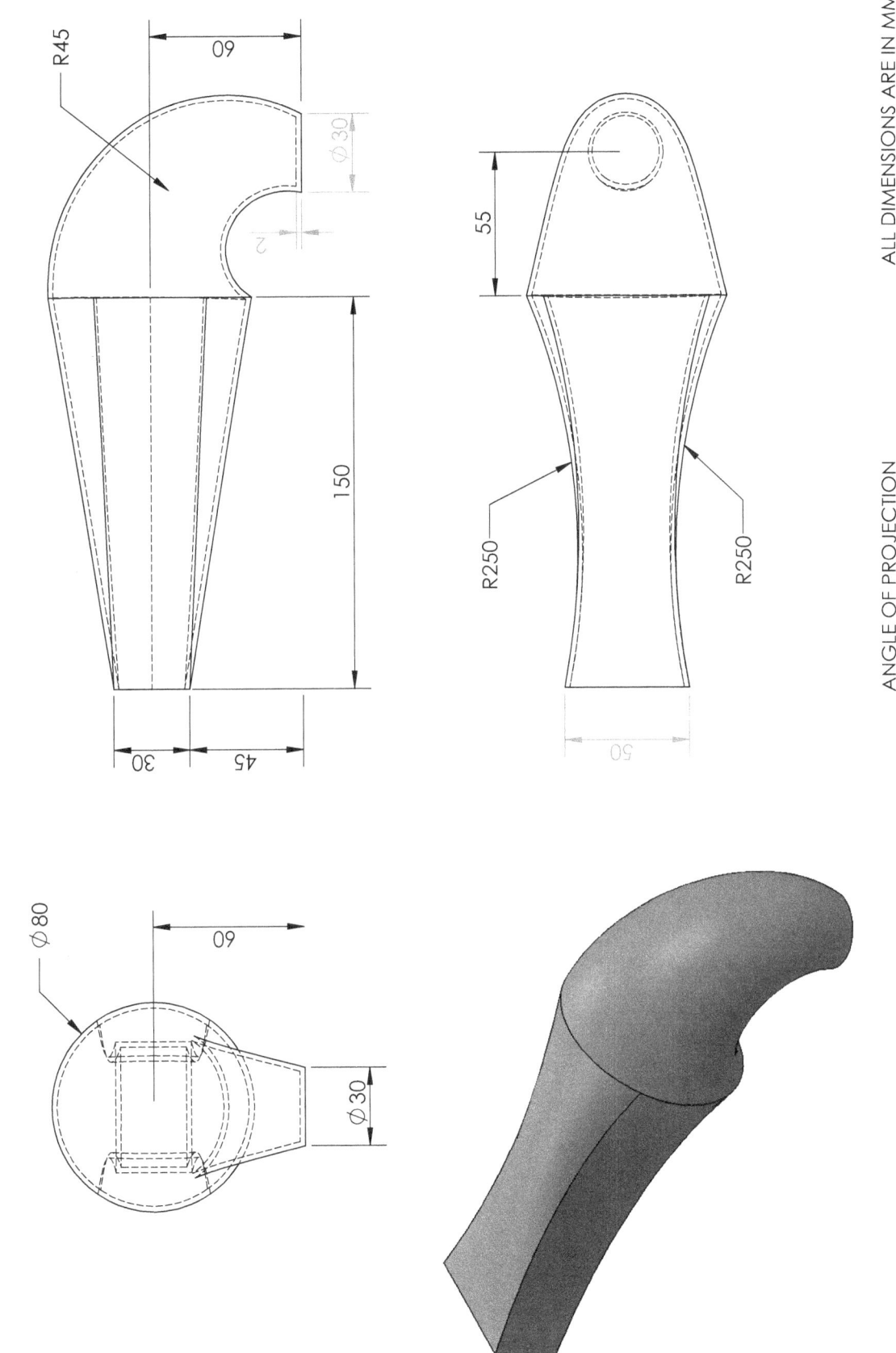

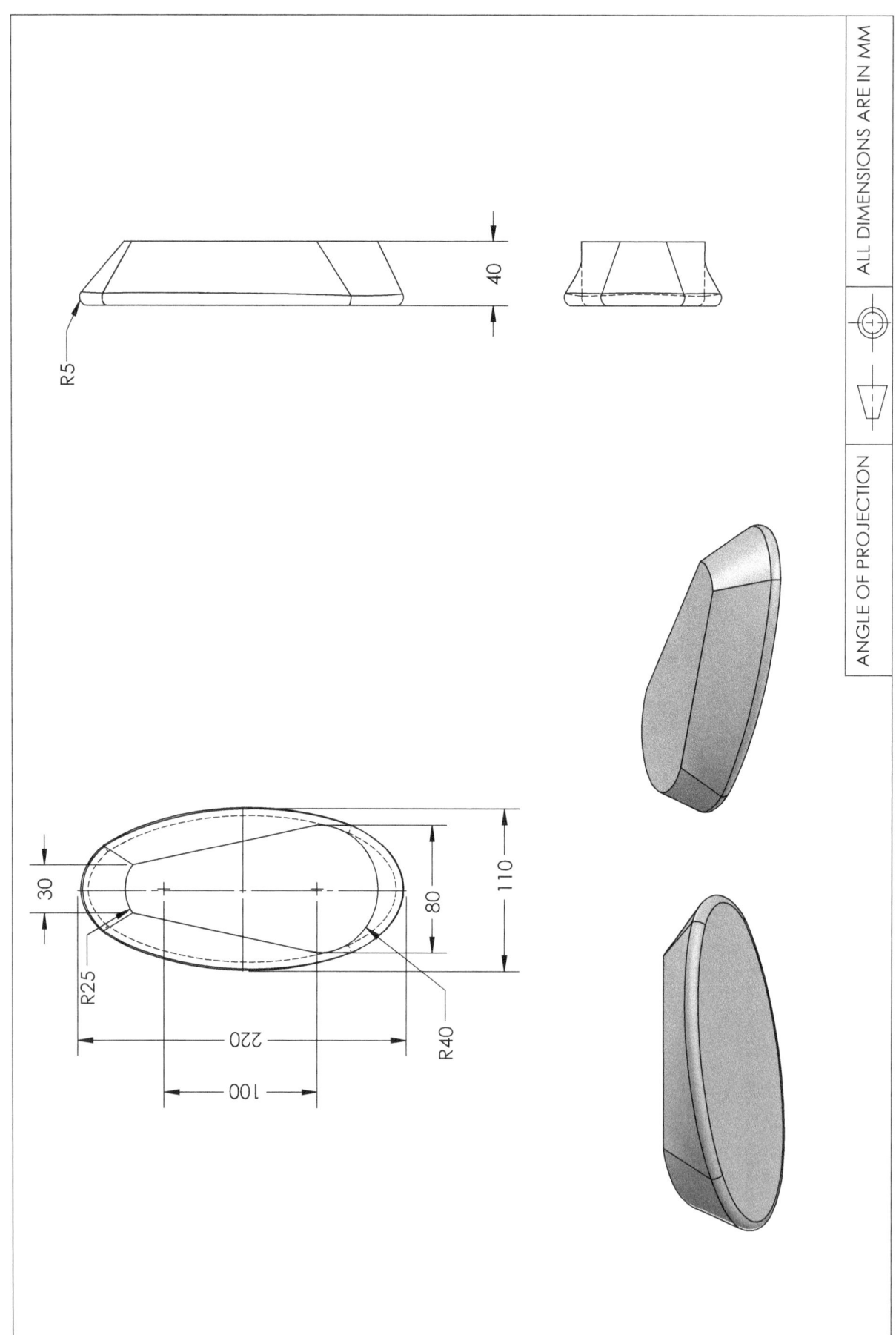

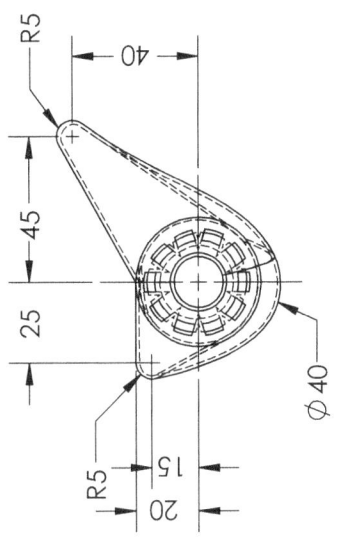
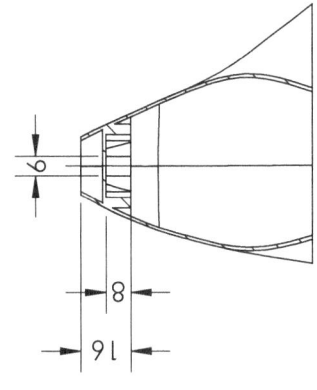
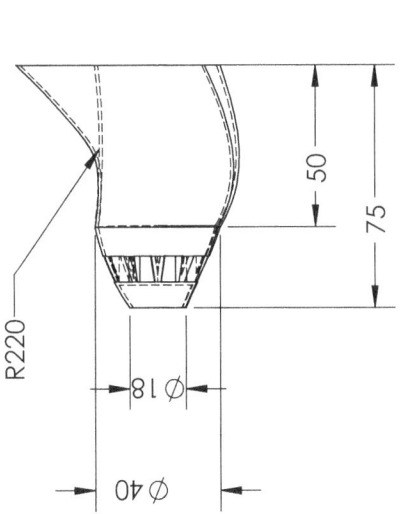
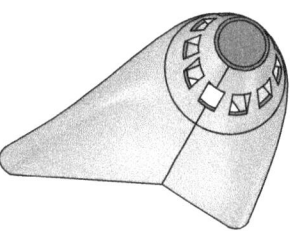
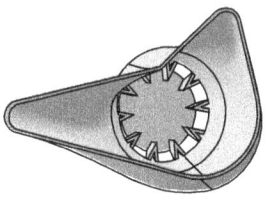

ALL DIMENSIONS ARE IN MM

ANGLE OF PROJECTION

Wishing you all
The Very Best.

www.ingramcontent.com/pod-product-compliance
Lightning Source LLC
Chambersburg PA
CBHW060415220526
45465CB00008B/2888